MW00441870

The Rise of the Silver Queen

Mining the American West

Series Editors

Duane A. Smith
Robert A. Trennert
Liping Zhu

The Rise of the Silver Queen

GEORGETOWN, COLORADO, 1859–1896

Liston E. Leyendecker, Christine A. Bradley, and Duane A. Smith

UNIVERSITY PRESS OF COLORADO

© 2005 by the University Press of Colorado

Published by the University Press of Colorado
5589 Arapahoe Avenue, Suite 206C
Boulder, Colorado 80303

All rights reserved
Printed in the United States of America

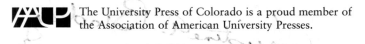 The University Press of Colorado is a proud member of
the Association of American University Presses.

The University Press of Colorado is a cooperative publishing enterprise supported, in part, by
Adams State College, Colorado State University, Fort Lewis College, Mesa State College,
Metropolitan State College of Denver, University of Colorado, University of Northern Colorado,
and Western State College of Colorado.

The paper used in this publication meets the minimum requirements of the American National
Standard for Information Sciences—Permanence of Paper for Printed Library Materials. ANSI
Z39.48-1992

Library of Congress Cataloging-in-Publication Data

Leyendecker, Liston E.
   The rise of the Silver Queen : Georgetown, Colorado, 1859–1896 / Liston E. Leyendecker,
Christine A. Bradley, and Duane A. Smith.
      p. cm. — (Mining the American West)
   Includes bibliographical references and index.
   ISBN 0-87081-793-0 (alk. paper) — ISBN 0-87081-794-9 (pbk. : alk. paper)
   1. Silver mines and mining—Colorado—Georgetown—History—19th century. 2. Gold mines
and mining—Colorado—Georgetown—History—19th century. 3. Georgetown (Colo.)
History—19th century. I. Bradley, Christine A. (Christine Ann), 1952– II. Smith, Duane A. III.
Title. IV. Series.
   F784.G34L495 2005
   978.8'02—dc22

                                                                                          2004029435

Design by Daniel Pratt

14   13   12   11   10   09   08   07   06   05   10   9   8   7   6   5   4   3   2   1

Contents

Acknowledgments

THIS IS LISTON'S BOOK, OR AT LEAST THE KIND OF BOOK HE ALWAYS WANTED TO WRITE after he retired from the History Department at Colorado State University. In January 2001 he sat in my office at the Clear Creek County Archives—up to his elbows in the papers of the Griffith Mining District—and told me I might have to finish the history of Georgetown. I muttered something like "you're not going to get out of this." Four months later I attended his funeral in Fort Collins with a lump in my throat and his words ringing in my ears.

The history of Georgetown needed to be written, and it needed to include the history of mining, engineering, and economics, set within the broader context of American history. I had a couple of thousand pages of research notes on the people, events, and buildings in town, but I knew I would flounder attempting to write much about mining and, gulp, economics. Fortunately, Ron Neely (president of Historic Georgetown, Inc.) called Duane Smith—Liston's longtime friend and colleague, an experienced author, and an expert on western mining and history. The publication was soon under way.

Liston left us an early draft of two chapters, his last publication, *The Griffith Family and the Founding of Georgetown* (University Press of Colorado, 2001), and "Georgetown, 1878" (Chapter 6 of this work). Duane and I developed the rest.

Harold Frost, longtime collector and expert on early Colorado photography, graciously allowed us the use of many photos. Thanks also to Liston's (and Duane's) former student and friend Lee Behrens, who shared his collections with us. The staff members of the Denver Public Library, Western History Department (especially James Jeffrey and Coi Drummond-Gehrig); the staff of the Colorado Historical Society Library (especially Rebecca Lintz and Barbara Dey); David Hays, archivist at Norlin Library, University of Colorado at Boulder; and Jessie Randall (no relation to the Jesse Randall of the nineteenth-century *Georgetown Courier*), special collections librarian at the Tutt Library, Colorado College, were invaluable in helping us retrieve photographs to illustrate this text. Eric Clements graciously offered to index the manuscript as a tribute to Liston.

Thanks to Matt Taylor's expertise as head of Clear Creek County's Mapping Department and Frank Young's knowledge of the general area, the text contains several maps to keep the reader oriented.

The Board of Directors and staff of Historic Georgetown, Inc., supported the project from the start. Liston's good friend and colleague Jim Hansen agreed to write the dedication, with our thanks and appreciation.

My personal thanks to Duane Smith for stepping up to the plate (in acknowledgment of his love of the national pastime and the Chicago Cubs) and sharing his knowledge of western mining history, as well as his expertise in writing and publishing manuscripts. Without his general direction (not to mention his chapters), I would be 300 pages into the town's history and only be up to 1870.

Special thanks to Barbara Leyendecker whose willing support and encouragement made this book possible. Duane's wife, Gay, always showed up with a smile and the ability to edit with great skill. My mother, Helen, provided constant encouragement. Ron Neely got the project started, read drafts, helped with illustrations, and was the best friend a project (and coauthor) could have. Ruth Rosenfeld came up with the title, and dear friend Barbara Page, now deceased, bequeathed funds to Historic Georgetown, Inc., which assisted with the final production and publication—something I know she would have liked.

CHRISTINE BRADLEY
GEORGETOWN, COLORADO, 2005

Dedication
Liston Edgington Leyendecker (1931–2001)

LISTON LEYENDECKER WAS BORN IN LAREDO, TEXAS, ON JANUARY 19, 1931. HIS father worked in the oil business, and his mother was an educated woman who instilled in her son the importance of being courteous and the satisfaction of intellectual curiosity. Liston's paternal family included Mexican ancestry, and he thrived on the influences of a multicultural upbringing. He valued equally the northern European and Hispanic elements of his heritage and, fully bilingual, relished opportunities to speak Spanish. Out of this background emerged a man uncommonly gracious and open-minded. He possessed a rare ability to make anyone—college student, businessman, or janitor—feel welcome and respected. He never forgot a name.

After graduating from the University of New Mexico in 1958, Liston moved to Colorado and eventually began graduate work in history at the University of Denver. He earned a master's degree in 1961, which included a thesis on Central City mining. He taught briefly at a Texas college and then returned to Colorado to pursue doctoral studies. Concurrently, he was hired by the Colorado Historical Society as deputy state historian and assistant editor. This affiliation honed Liston's appreciation for the way state and local history might be used to illuminate the past in practical ways. Although he left the society in

1966 to pursue an academic career, he later provided decades of invaluable volunteer service as a member of its Board of Trustees.

Following a two-year stint as History Department chair at Arapahoe Community College, Liston accepted a teaching appointment at Colorado State University (CSU). Thus began a remarkable thirty-one-year tenure during which he won awards for distinguished teaching and became a leader in awakening local Colorado communities to the importance of historic preservation and restoration.

Liston was a master teacher. Students enrolled in his Colorado and U.S. history courses experienced an instructor whose knowledge was solidly rooted in his own research. He also had a compelling knack for incorporating the perfect anecdote to illustrate a point. Especially notable, though, was Liston's commitment to making history useful. In 1976 he established a public history graduate program at CSU, which trained students by placing them in historic preservation and local history internships throughout Colorado. His carefully prepared interns ably assisted countless communities and agencies, providing outreach in the best sense of CSU's land-grant institutional mission. The students, meanwhile, gleaned invaluable practical experience, which subsequently enabled virtually all of them to find permanent employment as public historians. Liston was the best of mentors—someone who genuinely cared about each student. By example and, if necessary, by tactful nudging, he showed them that civility and integrity were qualities that enhanced one's life and contributed to success.

Liston was a fine scholar, blessed with the attributes most essential to this endeavor—curiosity, attention to detail, and perseverance. Historian and friend Duane Smith once said to him, "History is a mystery, and you are a great Sherlock Holmes." These characteristics were clearly reflected in Liston's quest to determine the truth of a story about George Pullman, a nineteenth-century businessman responsible for promoting railroad sleeping cars into a multimillion dollar personal fortune. According to legend, he had gotten the idea for pull-down berths in railroad cars by observing the fold-up cots on miners' cabin walls. An interesting story, but was it true? Ever curious, Liston decided to find out, and, in doing so, he embarked upon a research odyssey encompassing nearly two decades and culminating in the definitive biography of George Pullman. In the end, he learned that the story was a myth.

After retiring and even after being diagnosed with terminal cancer, Liston continued to conduct research on another lifelong scholarly passion: the history of Georgetown, Colorado. Most days he could be found at the Clear Creek County Archives or libraries in Denver or Fort Collins, working on this project. It was something he truly wanted to finish, and knowing that a favorite former student, Christine Bradley, and a respected fellow historian, Duane Smith, have completed his work would have given him enormous satisfaction.

Liston once described Georgetown as a place that's "as close to heaven as I'm going to get on this earth." In fact, this feeling prompted him and his wife,

Barbara, to buy a home there in 1996. My family spent many holidays with the Leyendeckers, friends whose company we valued and enjoyed. I especially remember a 1997 Christmas Eve in Georgetown. After a fine dinner at their house, Liston suggested that we take a walk. During the stroll I sensed what he meant in describing Georgetown as his heaven. The night was invigoratingly cold, and gently falling snow had covered everything. Our footprints were the first to mark the ground. Colorful Christmas lights from old Victorian-Era houses guided our way, and occasionally we smelled the smoke from wood-burning fireplaces. We seemed to be alone in what felt like another time and place. It was one of those indelible moments when you are fully aware of being alive. Or, for Liston, of being in heaven on earth.

The next morning we walked again, and Liston provided historical commentary on many of the residential and mining structures we passed. I realized that his research and carefully crafted reports had contributed to the survival and restoration of many of these properties. For him, history consisted of more than written records. It was also the homes where people lived, the business buildings where they earned a livelihood, and the mines where they toiled and sometimes died. The Hamill House, the Georgetown Loop Railroad, the Central City Opera House, and the Avery House and Old Town in Fort Collins all exist as they do in part because of Liston's dedication to preserving the tangible past.

Liston Leyendecker was much more than an exceptional professional historian. He was truly a person in whom the qualities of decorum and character were ideally combined. His essence is perhaps best captured in the Spanish language he so loved—*un gran caballero*. Those of us who knew Liston are grateful to Christine Bradley and Duane Smith for perpetuating his legacy.

JAMES E. HANSEN II
PROFESSOR EMERITUS OF HISTORY
COLORADO STATE UNIVERSITY

Prologue: Birth of a Legend

O accurst craving for gold!

—Roman poet Virgil

For lust and eagerness after gold and other things make men blind.

—Georgious Agricula

Miners will cross heaven to get to hell in search of gold and silver.

—Mining saying

Gold! Gold at Cherry Creek. Found all places prospected.

—*New York Times*, September 20, 1858

Testimony through the ages testifies to the lure of gold and silver on individuals' imaginations and actions. Exploration and settlement came to the New World because of those metals, which lured adventurers over mountain and desert. That "fever," as they called it, affected all nations, all peoples.

Despite great expectations, North America did not experience a gold rush until 1828–1830 in western Georgia and North Carolina. Then came the incredible news of the gold discovery in California in 1848 and the great rush to stake a fortune the next year. California changed American destiny and affected nearly every part of the globe. Nothing ever matched that rush for size, excitement, and significance.

Exactly ten years later more exciting news reached the States—a "second California" had been found in Pike's Peak country. Off went the adventuresome, the desperate, the naive, the hopeful, and the confidently expectant—along with a few experienced Californians—on a thirty-day journey from the Missouri River to the Rocky Mountains. The Mississippi Valley and the East were aflutter with golden expectations and worries, as clouds of sectional conflict boiled over the horizon.

At the same time, those on the West Coast heard news of silver strikes on the Comstock Lode. They trudged eastward over the Sierras to an inhospitable range of desert mountains to dig out a few silver fortunes that would far surpass anything found in the Rocky Mountains over the next few years. For the only time in American history, two major mineral rushes took place simultaneously. One marched westward, the other eastward; otherwise, expectations remained the same.

The Pike's Peak rush eventually gave birth to Georgetown far up Vasquez Creek, now called Clear Creek. What is about to unfold is the saga of this town, its people, and its mines—part of the larger story of the settlement of the Rocky Mountains and the trans-Mississippi West.

The Rise of the Silver Queen

1 Silver Threads Among the Gold

TALL TREES, BEAVER PONDS, AND STEEPLY PITCHED MOUNTAINS PRONE TO BURSTS OF strong wind greeted David and George Griffith in the summer of 1859 as they ventured up Clear Creek from Idaho Springs to the valley soon to be known as Georgetown. Like many who traveled to the goldfields of Colorado in 1858 and 1859, they had quickly wearied of the crowds of anxious greenhorns searching for mineral wealth around the early districts of Central City and Idaho. The brothers trekked off on their own to undeveloped lands soon to bear their name.

The riches of Colorado's mountains peeked out from the streams as flakes of gold worked their way down the rushing waters. These tiny specks became calling cards for the adventurous who longed to follow the creeks and valleys to their source in hopes of finding riches. Gold, silver, lead, copper, and other minerals lay buried within the hillsides, but early prospectors struck out for gold—shiny, bright, valuable, and relatively easy to work once unearthed. Silver ore held promise, but the cumbersome milling and processing requirements meant miners who came across the "argentiferous" ore generally made note of it only long enough to gather their energies and turn back to the search for gold.

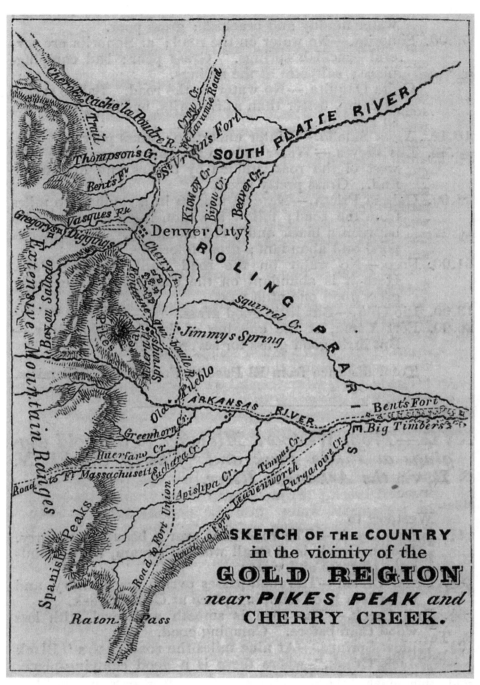

Map of Pike's Peak gold region. Photo from Marcy's, The Prairie Traveler, *1859. Courtesy, Denver Public Library, Western History Collection.*

The Griffith brothers, Kentuckians by birth, arrived in the Denver area from their home in Iowa as members of the Plattsmouth party, fifty-six men and one woman traveling across the plains in search of rumored riches in the fall of 1858.[1] The approach of winter meant most of these early settlers started by building shelter, gathering supplies, and occasionally panning local creeks while preparing to wait out the winter and position themselves to start mining early in 1859, ahead of the thousands anxiously waiting back in the States.

Gold mining in the mountains started in January 1859 with George Jackson's discoveries along Chicago Creek, just south of today's Idaho Springs. When Jackson publicized his discovery at the start of April, several hundred men went to "Jackson's mines," the diggings along Chicago Creek and the early Spanish Bar District along South Clear Creek (then known as Vasquez Fork).

Although these prospectors recovered some gold, the strikes were limited and not nearly as remunerative as they had hoped. When John H. Gregory, a teamster and former miner in the gold camps of his native Georgia, uncovered gold on May 6, 1859, on the side of a gulch that emptied into North Clear Creek, nearly all the miners deserted the South Fork for "Gregory's diggings." The find proved rich, but, not surprisingly, it did not make millionaires out of all the hopefuls who flocked to the area, causing many to continue to search for better opportunities.

Like many others, George and David Griffith arrived at the Gregory diggings too late to locate a claim in the bonanza area. They soon headed over the ridge to Spanish Bar, two miles west of the Jackson strike on Chicago Creek, before deciding to work their way upstream.

On or about August 1, 1859, the brothers came to a series of swampy beaver ponds nestled in a deep valley forked by two branches of South Clear Creek.[2] They bivouacked in the valley destined to become Georgetown. Using the camp as a base, they began searching the area for indications of gold. While David was off hunting, George discovered the gold-bearing Griffith Lead (pronounced "leed"), or quartz lode—an outcropping of promising ore—on the side of the mountain they named for themselves: Griffith Mountain.[3] Reports claimed George sluiced $100 in two days from his strike.[4]

Over the next several days the Griffiths uncovered the Turner and other lodes on the steep sides of Griffith Mountain and went on to bound and name the unofficial Griffith mining district.[5] George traveled to Central City to register his claims. He and David were soon joined by their brother John, whose arrival in the goldfields has been shrouded by the mists of history, with some reports stating he had also been in Denver in 1858.[6]

The three brothers built a cabin[7] in the valley floor and sank a twenty-foot shaft on the Griffith Lode from which they managed to extract about $500. They continued with a six-foot shaft on the nearby Turner Lode and a twenty-five-foot shaft to improve the Silver Lode, all in the summer of 1859.[8] The name "Silver" indicated the presence of silver in the district, but the brothers

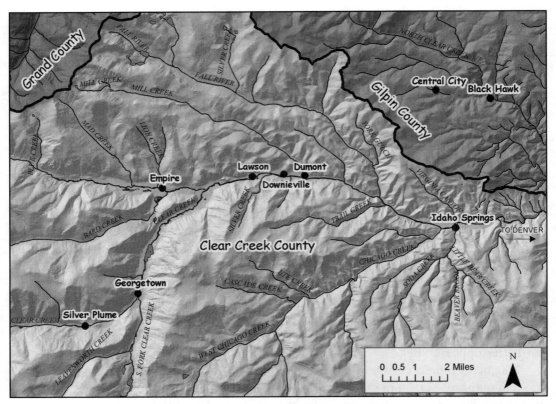

General map showing area from Denver west to Georgetown/Silver Plume, including Central City/Black Hawk. Prepared by Matt Taylor, Mapping Department, Clear Creek County.

never pursued the white metal. Some reports indicated that the Griffith discovery was the first silver mine in the territory. In his 1889 *History of the State of Colorado*, Frank Hall stated: "No one contests the assertion that George F. and his brother D. T. Griffith were the first to discover silver mines, near the spot where Georgetown was built."[9] The fact remains, however, that although they found silver, they made no effort to pursue its production.

The brothers, looking for even greater riches, continued to prospect around the region, even traveling north to investigate rumored strikes around Walden in the area known as North Park. They were not in camp when William Newton Byers of the *Rocky Mountain News* visited their Georgetown workings on August 26 and 27 to report these findings:

> Late in the evening [August 26, 1859] we reached the Georgetown mines, and were hospitably tendered the use of a house so long as we desired to stay, which was very acceptable after a long day's ride in the rain.
> August 27. We spent the forenoon in looking at the various mines, and prospecting. The first discovery was made by George Griffith some three or four weeks since. He opened his claim, which is on a lead high up on the

mountainside, set a sluice, and worked two days, taking out a little over one hundred dollars. Here the wall rocks closed in, presenting some difficulties not easily surmounted. About the same time, report came from the north of "thousand dollar diggings," or something like that, when Mr. G. up stakes, and off for new scenes, and his claim remains idle. Two or three other claims are being opened on the lead, but none are yet properly worked, although they evidently are very rich. Several other leads crop out in the immediate vicinity, on both sides of the creek presenting more favorable opportunities for working them, the creek cutting through at right angles, but there is no one to develop them; the whole settlement consisting of five men and one boy.

The creek is also rich in gold, but the bed rock is deep and it requires a combination of labor or capital to work it successfully.

We have no doubt there are millions of gold in the immediate vicinity.[10]

The northern part of the state failed to hold the Griffith brothers' interest for any length of time, however, and they soon returned to their Georgetown valley. That fall John Griffith returned to Glenwood, Iowa, to pick up his family. David went east to procure a stamp mill. George remained in the area, although not in the frosty confines of Georgetown.[11]

Certainly, the Griffith brothers were not the only prospectors to leave the goldfields with the onset of autumn's chilly evenings. Around 200 or 300 Pike's Peakers wintered in New Mexico, and part of their number planned to prospect not only that region but Arizona as well during the upcoming season.[12] Those who stayed in Colorado intended to work through the fall and early winter; however, most mines along South Clear Creek lay dormant until spring.

The people who remained in the Pike's Peak gold region during the winter and spring of 1859–1860 started to develop the infrastructure and improvements needed to support their efforts. They kept busy organizing new towns to provide supplies, building roads to allow equipment and supplies to reach the mountain camps, and pursuing whatever mining endeavors the weather would permit. Urban rivals Denver and Golden grew quickly, establishing themselves as entry and supply points for the various mining regions. New stores selling everything from mukluks to mining supplies appeared on crowded city streets. Roads to the goldfields began to develop as the need to transport heavy equipment to the goldfields became apparent. The need for roads quickly outweighed the funds available to build them, leaving many towns isolated with only dreams of a mining future.

Meanwhile, Georgetown lay blanketed in snow, awaiting the return of its founders and a handful of believers. As late as the outset of May 1860 the *Rocky Mountain News* reported that it had received no notices whatsoever from "the Georgetown diggings, on the extreme head waters of Clear creek." The editor surmised that since the camp was so close to the "Snowy Range" (the Continental Divide), it was doubtless still out of communication because of snow and ice.[13]

John Griffith returned to Denver in June 1860, accompanied by his wife, Elvira, his father, Jefferson—a "yonder sider," or veteran of the California goldfields—and a party of men. They rode through Central City to Eureka Gulch. The group spent several days attempting to reach Georgetown, only to discover that their bulky supplies could not be hauled across the mountains without a road.[14] This dilemma led them to obtain the right to build a toll road up Eureka Gulch from Central City over to York Gulch, then down Fall River to the valley of South Clear Creek.[15]

The only option was to build the road themselves, since no public funds existed to construct improvements of that nature.[16] They formed a toll road company to put in their course and the requisite bridges, expecting to reimburse their expenses by collecting tolls once they completed the project.

Work began on June 18, 1860, starting in Central City, building up Eureka Gulch across the ridgeline, then down toward Fall River. The workers found the residents of Iowa District (located along the banks of York Gulch and Fall River) eager to support a road, even a toll road; they signed and recorded a petition of support into the district's records on July 21, 1860.[17] Several days later the road reached the mouth of Fall River. The brothers decided to continue building west along the banks of Clear Creek.

Upon learning of the organization of a Fall River mining district, the Griffiths lost no time sending a delegation to the meeting, where they secured the district's consent to run their toll road through its proposed boundaries. They overcame a difficult point on the mountain immediately above Fall River by building a stout bridge across the creek, swollen with waters from the annual spring runoff. With this accomplished, they opened the way for traffic to the western end of the valley and development of their mines and fledgling town.[18]

George F. Griffith and Company employed between fifteen and twenty men to build the road a distance of twenty miles, at a cost of $1,500. During 1860 the revenue from tolls amounted to $151.50. From December 1860 until late March or early April 1861, the road remained open and free to all users, probably because of heavy snows. When the owners erected their tollgate in the spring of 1861, however, they were heartily abused for doing so. They charged $1.00 for the twenty-mile round-trip. A supporter of the road and the Griffiths noted that "they stand ready at any moment to sell out at cost to any or all the districts through which this road passes."[19] Later, George and John Griffith, undoubtedly fed up with the aggravation, disposed of their four-ninths interest in the road for $1,000.[20]

While construction of the road was starting and the Griffiths' supplies were held up in Central waiting for the chance to travel west, the residents of Georgetown moved to formalize boundaries for their own mining district, complete with laws and regulations. Establishment of a mining district gave a geographic definition to a region within which certain procedures would apply. The Griffith Mining District was officially established on June 25, 1860:

> Commencing at a high point of Rock on the West Side of Clear Creek Some half mile above the Junction of Said Streams, extending one mile, and a half

on each Side of Said Stream in breadth, and running up Clear Creek to the Junction of the Stream a short distance above the House of George F. Griffiths & Company. Thence up Said Streams three miles and extending in breadth one mile, and a half from the Banks of Each Stream.[21]

Formal preemption (filing) of lode mining claims started immediately. The town of Georgetown followed shortly, with dedication of a 640-acre town site on June 29, 1860. Over the next week the paperwork must have overwhelmed the district recorder, as the town's residents and their associates scrambled to locate mining properties, water rights, ranch or farming claims, and even one additional town site—unnamed—located only two days later (July 1) at the eastern end of the district at the foot of Douglas Mountain.[22] Over the next month most of the flat land between Georgetown and the eastern end of the Griffith District would be claimed for ranch and farming purposes. Eight parcels, roughly half a square mile each, traversed the length of the creek—with claim number 1 belonging to John Griffith, number 2 to John Surber, numbers 3 and 4 to James and Joseph Reynolds, number 5 to H. Wickoff (sold to George Griffith in January 1861), number 6 to Jefferson Griffith, and numbers 7 and 8 falling within the unnamed town site just mentioned. One additional ranch claim, the Fairview Ranch, appears to have been located north of the forks of the Main and South branches of Clear Creek.[23]

The Griffith family also preempted several sections of land south of the creek in the adjacent Montana Mining District, making the Griffiths some of the largest landholders in western Clear Creek County. Perhaps father Jefferson's experiences in the California goldfields gave him a good understanding of the value of ranch land and farmlands after the start of a mining boom. In any event, by the fall of 1860 the Griffiths owned substantial mining, ranching, and water rights along South Clear Creek.

David Griffith returned to Georgetown with a crude six-stamp mill in September 1860.[24] Shortly afterward, as district surveyor, David surveyed Georgetown. Thus by the fall of 1860 the Griffith Mining District was a small but bustling mining camp. Its progress indicated John Griffith's knowledge of the law, George Griffith's familiarity with the procedures necessary to establish a mining camp, and Jefferson Griffith's experiences in California. The legal framework and processing works were in place, yet the future of the young town would depend on the value and volume of the ore they would extract and the competition with other, newer camps, always ready to draw the anxious miner and investor to the new "El Dorado."

As 1860 drew to a close the entire valley, from Buckeye Bar below Idaho Springs to Georgetown and Empire to the west, seemed poised for a boom. In December a reporter making his way along Clear Creek described the creek and its surroundings, concluding that it was "second to none" as a Rocky Mountain mining region. A great deal of capital was invested, and an immense amount of work was projected for the coming season. Not only did the area possess

Topographic map from Empire Junction west to Bakerville showing general terrain, rivers, and towns. Prepared by Matt Taylor, Mapping Department, Clear Creek County.

numerous rich quartz lodes, but the large bars of sand in the bed of Clear Creek contained substantial quantities of gold. Such indications of wealth, he believed, would justify his prediction that within another season Clear Creek would equal Gregory District as a mining area as it attracted "business, improvements, and richness."[25]

Early mining districts developed laws to regulate everything related to mining: how to stake a lode claim, ranch, town lot, bar mining claim, water right, timber claim, and other forms of rights and privileges. Although the Griffith Mining District was technically part of the far western lands of Kansas Territory, no county or other governments were operating in the area. As a result, each new mining district had to provide for law enforcement (sheriff and judge), as well as for a recorder so all claims, deeds, and matters of importance would be meticulously filed, or recorded, with a designated official above petty squabbles and capable of keeping a complete and accurate listing to minimize disputes over claim locations and ownership.

George Griffith became the first recorder in the Griffith District. E. Y. Williams served as judge of the miners' court, and R. M. Barker was elected sheriff.

On February 28, 1861, the Territory of Colorado was established by the United States Congress. On November 1 the territorial legislature created seventeen counties. Georgetown was located in Clear Creek County, with the county seat in the town of Idaho.[26] These early counties soon took over some of the duties of the mining districts, since they provided for a sheriff, judge, and recorder; however, the individual rules and regulations relating to claim size stayed in force until superseded by later state and federal laws.

The framework of these early district governments reflected their residents' personalities. Some districts operated under casual rules, using as little legal framework as possible, whereas others developed page after page of detailed regulations. Smaller districts may have been settled by lone prospectors wanting to be left to their own devices. Several districts prohibited attorneys from their miners' courts. Others came west with an eye to starting larger companies, with investors who needed to be reassured that their funds were invested in stable, law-abiding communities capable of protecting their investments with the full force of law, if necessary. The Griffith District laws were extensive, extending beyond mining to include penalties for offenses such as taking "building timber" from the district and procedures for obtaining liens or mortgages.[27]

Between January and March 1861, Griffith District residents occupied themselves by reviewing and codifying laws and establishing more permanent boundaries to mesh with the surrounding districts, Union and Montana.[28] On March 2, 1861, James Burrell, David T. Griffith, and Henry Chapeze were appointed to a committee to codify and revise the district's laws, an effort to clarify their rights and interests under the new Territory of Colorado.[29]

On March 4, 1861, George F. Griffith, James Burrell, and W. S. Gibson met with Union District commissioners George L. Nichols and D. M. Stewart to establish formal boundaries and dividing lines between Union and Griffith districts. The date of the meeting coincided with the inauguration of Abraham Lincoln as president of the United States. In later years the local paper claimed the trip to Empire included a stop at the ridgeline, soon to be known as Union Pass, where the first political meeting in the district was held: "The Griffiths were southern sympathizers, Burrell was a Union man. He suggested holding a meeting to celebrate the seating of Lincoln as President. Burrell stripped the bark from a tree and made a record of the proceedings on a blazed square."[30] The name of the pass and the surrounding hillsides (Lincoln and Douglas mountains) would remain as a testament to the politics of 1860.

Throughout the spring of 1861 glowing reports in the press extolled the virtues of Griffith Mining District. With the completion of the road from Central, laws, and other refinements, the new area held great interest for prospectors. By that time seventy-five leads had been discovered, and those tested showed good promise. Prospectors arrived daily and seemed pleased with the area. The camp had forty permanent residents, thirty-five of whom had arrived since January 24 of that year. These people needed convenient trading facilities, and it was felt a man could do well by bringing in a "small stock of assorted goods."[31]

In April 1861 a reporter, writing from Georgetown, reinforced the optimistic views of the area. The snow was disappearing, and grass was starting to appear in the valleys. The weather had been good for several days, and hopes were that it would continue so miners and mill men could get busy. Prospects looked good, and predictions indicated enough quartz would be dug out to keep twenty mills occupied.[32]

The district lay in a beautiful setting on the South Fork of Clear Creek, covered with trees and permeated with streams, along with sandy bars good for placer mining and bottom land already green with grass. Georgetown was located on a lovely site, and its proprietors and inhabitants would be careful to protect its stately shade trees; the finely shaped cedars, junipers, and pines; and its charming grove of aspens.[33] Favorable geology meant the labor required to open and work the leads was insignificant compared with that in other camps. The Griffiths stated that they had begun to take "pay quartz" from the first five feet of their tunnel, and they forecast that fifteen men would average ten cords per day from that lead alone.

Several new leads had been uncovered, among them the Columbus. Its discoverers had handed a sample of quartz to a Central City assayer who pronounced it to be as fine as any, if not the most valuable, specimen he had analyzed. Such paeans led several miners to form a company to prospect, develop, and work on the Columbus Lode. By then eighty-two leads had been discovered, and between 2,000 and 3,000 claims had been filed on them.[34]

A *Rocky Mountain News* reporter spent April 24, 1861, investigating the Griffith District, accompanied by one of the Griffiths (probably John). The reporter was "astonished by the District's wealth in precious metals" and "consider[ed] it the richest of all the mining districts yet discovered." He examined the Griffith Lead very carefully, since it was the most extensively developed of any in the area. The surface quartz "paid handsomely" where huge masses of it lay on the surface of the mountainside. Pockets of pay dirt netted owners fifty cents to a dollar per pan, whereas quartz paid ten to fifteen cents per pound.

One shaft on the Griffith Lead was 25 feet deep with miners digging two tunnels, removing surface quartz, and cording it in large quantities. The Griffith brothers, who owned 1,400 feet on the lead, were responsible for this work. A reporter from the *Rocky Mountain News* quoted their father, "an intelligent gentleman," as stating that they would all be millionaires in good time.[35]

The Griffiths built a 1,000-foot chute from their discovery claim, close to the summit, down to the valley floor to enable them to use Davidson's Mill, the first completed in the district. The quartz, even surface rock, contained large quantities of iron pyrites and "undoubtedly was one of the . . . if not the richest, yet found in the Rockies." On the surface the Griffith Lode appeared "bold, unique and peculiarly striking, and the crevice after leaving the surface from a width of eighteen inches, increase[d] in width as one descended into it."[36]

Colonel William Davidson completed the first mill in the district, which began operations on April 24, 1861. The Griffith brothers were still working on their six-stamp mill, which had arrived in the camp the previous September. The brothers hoped to have their mill operable by May 5. In reality, the Griffiths did not complete the mill until August 1861, and its performance on numerous ores from area mines was so dismal that they ceased operations in September 1862.[37]

On February 7, 1861, Jefferson Griffith died suddenly in Mill City of unknown causes.[38] He was fifty-eight years old, well loved, and highly respected. Although his sons contemplated returning his body to Missouri for burial next to their mother, he was buried on one of his properties near Mill City, southwest of present-day Dumont.[39]

After Jefferson's death, the Griffith family began to drift away from their town. Rumors of rifts between the brothers started to surface. William Griffith pulled up stakes and had left the area by June 1861. John and Elvira sold their considerable properties to George and David in July 1861 before returning to Iowa.[40] Thus by the end of summer 1861 the two remaining Griffith brothers, David and George, owned much of the mining district they had originally discovered, as well as the faltering stamp mill. The brothers continued to participate in district affairs, David often acting as his brother's deputy.

Young Georgetown was almost 2,000 miles from Fort Sumter; nonetheless, the first shots fired in the Civil War on April 12, 1861, sent shock waves all the way to the Rockies, affecting life throughout the West. Friendly political differences turned into matters of life and death as the conflict continued to expand. The gold camps of Colorado became embroiled in the controversy; a prospector's state of birth became a cause for concern, and residents of the young mining camps started to distrust one another.[41]

In December 1862 David Griffith's interest turned from mining to the continuing war. After joining the Third Regiment of Colorado Volunteers, David gave George his power of attorney and left the area.[42] David's support of the Union Army may have created some of the problems between the Griffith brothers. Certainly, George was viewed as a southern sympathizer, although the opinions of the other brothers are unknown. Nevertheless, by the end of 1862 George was the only member of the Griffith family who remained in the Georgetown area.

The closure of the Griffith Mill in September 1862 epitomized the problems of the early gold camp days in Georgetown: high hopes were made difficult by inferior technology. As Georgetown faltered, attention started to shift to the other side of the mountain. The town of Empire, focal point of the neighboring Union Mining District, boomed in June 1862 with the discovery of the wealthy Tenth Legion Lode. D. P. Dodd, John H. Adams, and James C. Huff located the valuable claim on Silver Mountain, northeast of Empire, on June 19. In August, *Colorado Mining Life* reported on a bustling scene: "Bully for Empire!—We hope all her citizens make their piles quicker than no time."[43]

Prospectors covered every inch of the district. Empire became the shining light of Clear Creek County. The rave reviews continued into September:

> Union is the great summum bonum of the Clear Creek country. Here some 250 hardy industrious sons of toil, labor for present need with every prospect of a big thing in their future. Here in a deep vale, shut in by alpine hills, at the base of the every [sic] snowy Cordillera, possessed of a splendid climate and every requisite for future greatness and prosperity, the future dense, Rocky Mountain metropolis, Empire City, is nestled in its infancy. . . . The town now contains some 70 buildings, and is rapidly growing. About a mile to the north is the site of the mining interests. About 150 men are engaged in sluicing dirt from the patch while a large number of others are at work on the leads above the placers.[44]

Georgetown was quickly overshadowed, limping along with somewhere between five and ten buildings, a handful of residents, and two struggling, ineffective, and nonproductive mills.

In October the same year the discovery of the Aldebarren Lode, later known as the Great Equator, north of Empire on Lion Creek, created a second wave of excitement. The discoverers, George L. Nickolls and Charles H. Martin, were quickly joined by other individuals well-known in Colorado mining circles: William H. Russell, Benjamin F. Pine, Henry Sherwood, Robert W. Steele, and John Peck. The success of these two major producers continued the frenzy of development in Empire Valley.[45] Georgetown could not compete with the newest El Dorado.

Georgetown's faltering mills and the inability to extract enough quality ore must have discouraged George Griffith as he watched his family leave for other parts of the country and his fellow miners head to Union District. Empire became the town he had envisioned for his valley. The badly needed attention of the territorial press was focused on Empire, with few if any mentions of the mines around Georgetown.

The winter of 1862–1863 was a slow time for the little Georgetown District. John Dumont, an 1858 pioneer and one of the few residents at the time, later wrote:

> I was a sojourner in that neck of the woods in the winter of 1862–3, and with George Griffith (the founder of Georgetown and named for him, who located there some years previous), David Griffith, H. K. Pierson and Almon B. Fellows were the only residents that winter. Edgar Freeman, Henry Chapeze, John T. Harris and others came later. The first cash money received was from and through Mr. S. F. Nuckolls, by the sale of the Griffith, Mills County and other mines to Eastern capitalists, resulting in the formation of several mining companies, one being the What Cheer' company of Providence, Rhode Island, who operated the Griffith mine under my management. A mill was erected under the supervision of John T. Harris, and named the What Cheer Mill. Mr. Wm. Hale of Providence, was the managing director.[46]

As Dumont mentioned, the first real cash to be invested in the Griffith District came through S. F. Nuckolls, a Central City businessman and associate of the Griffiths. On January 16, 1863, George and David (in absentia) contracted to sell their holdings to Stephen F. Nuckolls for $10,000.[47] In November they sold other properties to Nuckolls for $26,000. A month later George gave Nuckolls his power of attorney to sell various claims for him, an indication that he would soon leave the district bearing his name.[48]

Nuckolls brought new life and enthusiasm to the Georgetown mining ventures. After he took over the Griffith Mill he made general repairs on it and substituted new iron stamps for the old wooden ones. He restarted the old mill in June 1863 and ran it for four months on ores from the Griffith, Burrell, and Corasina lodes—recovering about $2,500.[49]

In the fall of 1863 Colorado's gold mines became popular with East Coast investors as William H. Russell, mastermind of the short-lived Pony Express and president of the freighting firm Russell, Majors and Waddell, headed to New York to sell stock in his Empire mines to eager stockholders. Investors flocked to the new industry—shares in Colorado's gold mines soared in value.

Stephen Nuckolls and many others followed Russell's lead. Nuckolls formed several companies with syndicates of eastern investors: the Wilson and Cass, Washington Mining Association, the Mount Alpine Company, and the Georgetown Gold Mining Company. Georgetown's potential improved with the infusion of cash. The new companies, funded with East Coast dollars, built mills with the newest technology in one final attempt to unlock the secret to successfully developing Georgetown's golden treasures.

The Wilson and Cass Company constructed a mill and dam, named Plummer's Dam after its manager J. E. Plummer, at the north end of the town site. The company's account book shows company investors provided $1,000 a month to pay expenses, including Plummer's monthly salary of $250 and the salaries of his employees—many of whom were simply referred to as "Cornish miners." Considerable money also went to services for roadwork, probably indicating the need for serious repairs and improvements to the early Griffith Road from Central City.[50]

The Washington Mining Association developed at the base of Griffith Mountain on the east side of town, close to the original Griffith mining properties (the site later known as the Capital Prize Mine). The Mount Alpine Company built a mill at the foot of newly named Alpine Mountain, along the Leavenworth Branch of Clear Creek where the present-day hydroelectric plant stands. Although it was constructed as a gold mill, its superintendent, Colonel William Hale, soon began to look to silver as the possible future for his property.

These corporate efforts provided an influx of cash and jobs for Georgetown, yet they never generated enough revenue to pay their operating costs, much less to turn a profit. As the days of the individual prospector began to fade, requirements for working capital, engineering knowledge, and metallurgical skill came to the fore. Even with money to fund the continual parade of new technologies, profits remained elusive.

The investment bubble burst in the spring of 1864, as stockholders wearied of the endless dollars needed to extract minerals from Colorado's mountains. Major investors pulled their funds from towns such as Empire, leaving the companies that had just started to invest in Georgetown in a state of confusion. George Griffith left town shortly thereafter, traveling to Durango, Mexico, on behalf of Stephen Nuckolls and others. He moved to Mexico as superintendent of a gold mine just as his namesake town was about to give up on gold and turn to a new mineral—silver—for its livelihood and prosperity.[51]

Georgetown's "golden" era came to an end with the departure of the Griffiths. As happened in many early mining camps, the great hopes and dreams faltered as the realities of processing and investment replaced the excitement of prospecting and finding "free" gold in the creek beds. Fifteen years later a local reporter summed up the era:

> [I]t was a sickly and despondent camp for a long time. The stamp mill could not save much gold because there was but little gold in the quartz.
> Provisions were high. It was all outgo and no income. However, game was abundant, and the beaver ponds were full of delicious trout. It is enough to create a famine in one's stomach to hear such old timers as H. K. Pierson and John T. Harris tell about the abundance of game and fish in those early days, when the foundations of the now-prosperous "Silver Queen" were being securely laid.

Mining for gold did not pay because the gold was not in the quartz. It was beyond the metallurgical skill of human beings to make gold mines out of silver mines.[52]

NOTES

1. Liston Leyendecker, *Georgetown: Colorado's Silver Queen 1859–1876* (Fort Collins, Colo.: Centennial, 1977), 7. The Griffiths arrived with groups from Missouri, eastern Kansas, and southeastern Nebraska. See also Nolie Mumey, ed., *Anselm Holcomb Barker, 1822–1895; Pioneer Builder and Early Settler of Auraria; His Diary of 1858 From Plattsmouth, Nebraska Territory, to Cherry Creek Diggings, the Present Site of Denver, Colorado* (Denver: Golden Bell, 1959).

2. Two dates are commonly used for the Griffith brothers' arrival in what would become Georgetown: August 1 and June 15. William N. Byers of the *Rocky Mountain News* reports on the district in late August, indicating it had been in existence for four or five weeks, whereas a retrospective in the *Colorado Miner* (January 13, 1877, 2) says "on or about the 15th of June" and gives the exact date of discovery of the Griffith Lode as June 17, 1859. The most common reference appears to be "on or about August 1."

3. *Georgetown Courier,* March 27, 1884, 3.

4. Ovando J. Hollister, *The Mines of Colorado* (Springfield, Mass.: Samuel Bowles, 1867), 71; see also Frank Hall, *History of the State of Colorado* (Chicago: Blakely, 1889), I: 227.

5. *Rocky Mountain News,* April 10, 1861, 1; Hollister, *Mines of Colorado,* 71.

6. Benjamin Draper, "Feudin' and Fightin': 1860," *Rocky Mountain News,* June 6, 1948, A-12; Mumey, *Anselm Holcomb Barker,* 61.

7. In later years the editors of the *Georgetown Courier* placed the early cabin site on property then owned by L. W. Berry who owned the property along Fourteenth Street between Griffith and Main.

8. The *Mining Review* [Georgetown], January 1873, 3–5; *Rocky Mountain News* (weekly), April 10, 1861, 1.

9. Hall, *History of the State of Colorado,* I: 317.

10. *Rocky Mountain News* (weekly), September 10, 1859, 2.

11. *Rocky Mountain News,* May 2, 1861, 2.

12. *Rocky Mountain News,* November 10, 1859, 2.

13. *Rocky Mountain News,* May 9, 1860, 2.

14. *Rocky Mountain News,* May 2, 1861, 2.

15. Ibid.

16. LeRoy R. Hafen, ed., *Colorado and Its People: A Narrative and Topical History of the Centennial State,* 4 vols. (New York: Lewis Historical Publishing, 1948), I: 184–185.

17. Iowa District, Book A, 72–73, Clear Creek County Archives, Georgetown, Colorado.

18. *Rocky Mountain News,* May 2, 1861, 2.

19. *Daily Evening News,* April 12, 1861, 2.

20. *Rocky Mountain News,* May 2, 1861, 2.

21. Liston E. Leyendecker, *The Griffith Family and the Founding of Georgetown* (Boulder: University Press of Colorado, 2001), 48–49.

22. Griffith Mining District, Book A, 102, Clear Creek County Archives, Georgetown, Colorado.

23. Ibid.

24. *Georgetown Courier*, March 27, 1884, 3.

25. *Rocky Mountain Herald* [Denver], December 22, 1860, 4.

26. Hall, *History of the State of Colorado*, III: 310. Hall states that the first county seat was "Idaho Bar," which later became Idaho, then Idaho Springs.

27. For the full text of the laws of the Griffith District, see Leyendecker, *Griffith Family*.

28. Griffith Mining District, Book A, 9–14, 128–131, Clear Creek County Archives, Georgetown, Colorado.

29. Griffith District, "Minutes and Laws," 4–5, Clear Creek County Archives, Georgetown, Colorado.

30. *Georgetown Courier*, August 10, 1901, 4.

31. *Rocky Mountain News* (weekly), April 10, 1861, 1.

32. Ibid.; *Rocky Mountain News*, April 27, 1861, 2.

33. *Rocky Mountain News*, April 27, 1861, 2.

34. Ibid. The original discoverer would file his or her right to the lead, then others would stake claims running in all directions from the original discovery; thus eighty-two leads could result in over 1,000 claims.

35. *Rocky Mountain News*, May 4, 1861, 2.

36. Ibid.

37. Ibid.; see also, *Georgetown Courier*, March 27, 1884, 3.

38. *Georgetown Courier*, December 31, 1904, 4.

39. *Rocky Mountain News*, May 8, 1861, 2.

40. Ezra B. and Sarah L. of Mills County, Iowa, were heirs of Jefferson Griffith, deceased. Book K, 474, Clear Creek County Archives, Georgetown, Colorado. Nannie or N. J. or Nancy J. Griffith followed suit and disposed of her properties as late as 1864–1865. See Book C, 37; Book K, 471; Book P, 145, all in the Clear Creek County Archives, Georgetown, Colorado.

41. See Duane A. Smith, *The Birth of Colorado, a Civil War Perspective* (Norman: University of Oklahoma Press, 1989), for further discussion of the Civil War in Colorado history.

42. David T. Griffith, Third Regt. Colorado Volunteers, to George F. Griffith, Power of Attorney, dated December 13, 1863, Book C, 625, Clear Creek County Archives, Georgetown, Colorado. The date should be 1862, as the transaction was recorded on November 9, 1863, but the notary public dated it December 15, 1862. George probably did not get around to recording it for a year.

43. Quoted in Louise Harrison, *Empire and the Berthoud Pass* (Denver: Big Mountain, 1964), 93.

44. Ibid., 94.

45. Ibid., 106.

46. *Georgetown Courier*, December 31, 1904, 1.

47. Book D, 43, Clear Creek County Archives, Georgetown, Colorado.

48. Ibid., 226.

49. *Georgetown Courier*, March 27, 1884, 3.

50. Xerographic copy of Wilson and Cass account book on file at Clear Creek County Archives, Georgetown, Colorado.

51. Leyendecker, *Griffith Family*, 44–45.

52. *Colorado Miner*, January 12, 1878 (special edition), 4.

Mining Districts

When throngs of miners descended on an area in search of minerals, the potential for disputed claims, claim jumping, and other forms of skullduggery quickly surfaced. To keep affairs in order, early settlements quickly turned to the framework of a mining district: a simple government created by the miners to give them the ability to establish and enforce laws relating to mining claims, timber, recording of documents, and other items essential to their daily lives.

Early Colorado districts (even those that pre-dated the establishment of the Territory of Colorado in 1861) undoubtedly took their structure from earlier camps in California, most of which were based on early Spanish mining laws. The miners created a government simply by calling themselves into session, establishing boundaries, writing bylaws, and electing officers. Each district required a recorder to make careful records of filed claims and keep the records safe from tampering or loss, a sheriff to make sure things remained peaceful, and a president or judge of the miners' court to call the miners' meetings to order.

The lands around Georgetown all fell within the territory of the Louisiana Purchase. The federal government owned the property but lacked the ability to manage the huge expanse of land. Miners and other settlers raced out in search of riches, leaving the details of ownership to later sessions of Congress.

In later years mining laws, homestead regulations, and other lengthy regulations settled into place. In the late 1850s prospectors moved west quickly—ready to stake their own claims, work their mines, and acquire riches with little concern as to ownership of the soil.

Nevertheless, even the most independent prospectors realized that mining districts were a necessity and the best means of protecting their interests.

Roads

Early miners often found their efforts blocked by a lack of access. Colorado's mountains and the valleys Mother Nature had carved through the ages often lacked the flat stretches required to transport wagonloads of supplies and the heavy equipment needed to process ore.

Roads provided economic lifelines to mountain mining camps, yet they also created numerous battles over rights-of-way. Everyone wanted the road unless it was built too close to their cabins or across their newly fenced pastureland. They wanted access *to* but not *through* their property. Furthermore, it was difficult enough to find the funds to build and maintain the roads. No one was going to take the time and money to survey miles of roads that traversed acres of public land.

The federal government wanted to encourage the development of its western lands and did not want to spend time settling disputes relating to roads and rights-of-way. The 1866 mining law (later included within section 2477 of the Federal Code) opened the way for roads to be built across the public domain: "The right-of-way for the construction of highways over public lands, not reserved for public uses, is hereby granted."

The message was simple, straightforward, and direct: if you build it, it is a road. You cannot build across private land without permission, but public land is open and ready for any roadway.

More than a hundred years later, Colorado's counties are faced with attempting to define rights-of-way that were built without survey or specific application, often the basis of the county road systems built in the 1860s with no documentation. The nineteenth century's shortcut has become the twenty-first century's nightmare.

Empire Pass, also known as Union Pass, looking south toward Georgetown. Courtesy, Harold Frost Collection, Alex Martin photo.

2 "The Maddest, Merriest Time Ever"

THREE EXPERIENCED PROSPECTORS HIKED THE HIGH PEAKS AROUND THE CONTINENTAL Divide in the cool fall days of 1864. Whether it was rumors of ore in Summit County's Peru Creek Basin, a general belief that the most valuable ores were nestled at high altitude (based on the rich silver ore found atop Cerro Rico in Potosi, Bolivia), or the simple beauty of aspens in the fall that had attracted them to the region, they succeeded in finding extremely rich silver ore on or about September 15, 1864.[1] The Gus Belmont Lode was unearthed approximately eight miles south of Georgetown on the precipitous ridge of Mount McClellan at an altitude of 13,200 feet. The three gentlemen— Robert Steele (first and only governor of the extralegal Jefferson Territory), James Huff, and Robert Taylor, prospectors living in Empire—brought back ore samples, "blossoms," rich enough to start Colorado's first major silver rush.[2] Huff and Steele had been key players at the start of Empire's boom period and brought a reputation for sound knowledge of mines and mining.

The presence of silver in Colorado's mountains had been acknowledged for some time. Prospectors found silver ore at several locations in the territory: Georgetown and Buckskin Joe in 1859, Silver Mountain above Empire in 1860, and other early finds in Summit County, among others. The impact of the

Belmont claim came from the exceptional quality of the ore and the general economic conditions in the fall of 1864. Frank Hall later summed it up in his *History of Colorado*: "Gilpin County had sold all her better mines in New York, and the new owners had made lamentable failures of them. This [Belmont claim] was about the only productive region we had. Gulch and placer mining had reached the final stage, hence the revival of interest at the head of Clear Creek created a mighty sensation."[3]

The approaching end of the Civil War allowed East Coast and overseas investors to focus on new business opportunities. Interest in many of Colorado's gold mining camps waned as placer operations dried up, leaving only the more expensive and troublesome lode claims. By the summer of 1864 gold mining had gone from a get-rich-quick opportunity to a complicated science requiring sound management and extensive knowledge of mining and geology. With this reality, coupled with problems in the East Coast economy (particularly the stock market), Colorado's gold mining communities such as Empire began to struggle. Many of the state's better-known prospectors began to look elsewhere.

James Huff was the first to unearth the riches of the Belmont Lode, extracting samples of ore to a depth of two feet. Huff had been involved in the discovery of the Ida silver mine north of Empire in 1860, and he readily recognized silver ore.[4] As a colocator of the valuable Tenth Legion Lode in 1862, he also had experience with an extraordinarily valuable mining property. His efforts to locate a pass between Empire and Middle Park (later known as Berthoud Pass) in 1860 gave him excellent credentials as someone who enjoyed hiking the highest points along the drainages of Clear Creek. Robert Steele's involvement with the Aldebarren Lode provided him with similar expertise.

Believing they had unearthed the entrance to a small fortune, Huff and the others packed up their specimens and headed to Central City to see if the ore would assay as well as they expected. They sought the opinion and advice of Professor Frank Dibben, a respected assayer. Dibben's preliminary work put a value of $200 to $500 per ton on the samples,[5] impressing him so much that he soon left his office and joined forces with the three prospectors.

With valuable ore in their pockets and good times looming, the savvy prospectors, their assayer, and a select few others developed a strategy to secure their interests and acquire rights to as much land and mineral property as possible before others became aware of the district's richness. They needed enough partners to provide the necessary capital and stake substantial claims without diluting individual returns. Willing investors were quietly solicited out of Central City and Empire. Nine shares were divided as follows: William Slaughter and James Huff, John W. Peck and Robert W. Taylor, Jonathan Cox, Frank Dibben, A. S. Patterson, B. F. Pine, D. C. Vance, R. W. Steele, and George W. Lane.[6]

After stocking up with provisions at Peck and Patterson's Store in Empire, a party of seven started for the new district on September 29, 1864.[7] As with

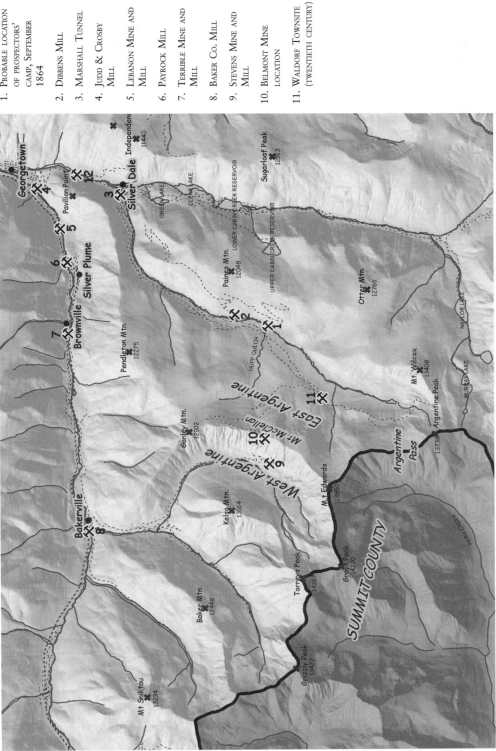

1. Probable location of prospectors' camp, September 1864
2. Dibbens Mill
3. Marshall Tunnel
4. Judd & Crosby Mill
5. Lebanon Mine and Mill
6. Payrock Mill
7. Terrible Mine and Mill
8. Baker Co. Mill
9. Stevens Mine and Mill
10. Belmont Mine location
11. Waldorf Townsite (twentieth century)

Topographic map of Georgetown to East Argentine with landmarks and mill sites marked and identified. Prepared by Matt Taylor, Mapping Department, Clear Creek County.

Dibben, both Empire store owners (Peck and Patterson) temporarily closed shop and headed out for the new discovery. The group headed southeast out of Georgetown on the Nan Smith Trail along the steep, craggy face of Alpine Mountain—crossing the creek at what was later developed as the Equator Tunnel. At the first major fork of the South Branch of South Clear Creek they turned southwest, following the Leavenworth Branch of the creek toward the Snowy Range.

The new entrepreneurs established camp at the base of the newly named Huff Gulch, then began to stake their claims to the hillsides, creeks, and open lands of the beautiful, remote basin. The thirtieth was spent prospecting. Formal actions took place on October 1. First came the establishment of the Argentine Mining District:

> Commencing at a point on the Divide between the waters of the "West Branch" and "Middle Clear Creek," where the western line of the Georgetown District line crosses said divide, thence following said line in a southerly direction to the summit of the divide between the waters of the Platte and Clear Creek, thence following said Divide in a Northwesterly Direction to the summit of the Snowy Range or the western boundary of Clear Creek County, thence northwardly along said County line to a point where said divide between said Middle Clear Creek and West Branch intersects said Range or county line, thence Eastwardly down said divide to the place of Beginning.[8]

Jonathan Cox served as president of the district, with Robert W. Steele as secretary. Subsequent assays of the ore from these early claims averaged $827.48 per ton. The future of the Argentine District shone bright.[9]

Robert Steele named the mountains in the valley. A staunch Democrat, Steele surrounded the new mining district with names from the upcoming election: "McClellan," the largest mountain, after 1864 presidential candidate General George McClellan and "Pendleton," the smaller peak to the east, after George H. Pendleton, the vice presidential candidate. The Belmont claim would also have been politically appropriate, since New York financier Gus Belmont was chairman of the Democratic National Committee during the 1864 elections.

Like the pioneers in Hollywood movies, these early adventurers must have swept their arms around them, claiming everything in sight. They started with three lode claims (Gus Belmont, Nominee, and Halifax), with eight additional claims to the northeast and the southwest. Next came three sets of tunnel claims: first the McClellan Mountain tunnel claims running from the foot of the mountain in Stevens Gulch to the northeast, with nine adjoining tunnel claims; then the Pendleton Mountain tunnel on the southeastern slope of Pendleton Mountain, running northwest, with three adjacent claims; and the third, the Jones Mountain tunnel, running northwest along with two adjacent claims.[10] Next came a series of water claims (1,000 feet each) starting at the mouth of

Huff Gulch and running 5,000 feet down and 4,000 feet up along Leavenworth Creek and 9,000 feet up Huff Gulch.[11] The final step was the recording of claims for 1,490 acres for "Ranch and Town purposes"—nine claims of 160 acres each, "[c]ommencing at the mouth of Huff Gulch and running one half mile South and one mile North thence west from the extreme ends of this line sufficiently for to embrace fourteen hundred and forty acres."[12] By the end of the first meeting of the Argentine District, big chunks of the eastern slope of Mount McClellan and most of the flatland and waters within the large valley had been secured.

That night sixteen inches of snow blanketed the area, ending the adventures of 1864. Caching their supplies for the winter, the party returned to Idaho Springs to record their actions with Clear Creek County. October 5 and 6 were busy days in the office of County Clerk and Recorder P. E. Charruaud and his deputy, Mr. Davis.

Ten years later an interview with Steele and Cox, printed in the *Colorado Miner*, revealed the nature and cost of the supplies the prospectors carried with them:

> One sack flour, $24; 20 lbs. sugar at 68 cts. per lb.; 2 lbs. tea at $3 per lb.; 3 lbs. dried apples at 35 cts. per lb.; 5 lbs. salt at 25 cts. per lb.; 2 lbs. lard at 40 cts. per lb.; 54 lbs. ham at 45 cts per lb.; 22¼ lbs. bacon at 35 cts. per lb.; 1 quart whisky, $1.50; 31½ lbs. potatoes at 15 cts. per lb. Besides the above, there were a number of tools and minor articles of grub, the whole bill amounting to $95.31.

The editor found himself amused at the list: "We don't know which will strike the reader of today as the most surprising—the high prices of provisions in these days, or the fact that these hardy pioneers started out into the wilderness with no coffee and only one quart of whisky for seven men!"[13]

Anticipation rose during the winter of 1864–1865. The wildfires of excitement and rumors of riches were fueled by the inaccessibility of the Argentine Basin. By the time the snow started to melt the following June, prospectors had spent almost nine months dreaming of wealth. County records show the first burst of new claims being staked on June 14. Prospecting parties went wild. One early miner later put the situation into perspective: "In the year 1865 the grand silver spasm broke out. It was the maddest, merriest time that Colorado ever saw."[14]

The county records swelled with lode claims on almost every inch of the Argentine District. Groups of prospectors from Empire, Central City, and other parts of Colorado Territory flocked to the new region. The well-known prospected alongside those just starting out. Henry and Willard Teller, Bela Buell, Joseph Watson, and others from Central City soon realized the potential of silver mining in Colorado. The first claims focused on the vicinity of the Belmont. Next, the prospectors headed east across Huff Gulch. Attention soon turned to the steep western slope of Mount McClellan. As Huff and the others had noticed, the veins tucked beneath the summit of Mount McClellan could be accessed from either side.

Travel up the South Branch of Clear Creek to Quail Creek (also known as Stevens Gulch) traversed wider canyons, meaning transportation and road building might be easier. The area would be divided into two regions: East Argentine, the site of the original discoveries, later known as the Waldorf Basin; and West Argentine, running from Bakerville to the valley formed by McClellan, Grays, and Irwins (later renamed Torreys) peaks.[15]

One of the first major claims in West Argentine was the Baker Lode on Kelso Mountain, located by James Baker, R. W. Davis, F. F. Brune, Richard Irwin, W. F. (Fletch) Kelso, P. E. Charruaud, A. M. Noxon, and others. One of the better-known tales about the area indicated that the prospecting party had started in East Argentine, found it too crowded, and scampered down the western slope of McClellan Mountain to try their luck in a quieter place.[16] Whether the story is fact or fiction, this party of Idaho Springs miners did open the second major silver mining district in the county and soon found themselves surrounded by the same crowds they had left on the other side of the mountain. It comes as no surprise that R. W. Davis, the hardworking deputy clerk and recorder whose notes recorded all the claims of the Steele party, and his boss, P. E. Charruaud, were among the first to head to the new district with the melting of the winter snows.

Ironically, the early prospectors who raced to the Argentine District walked right over the hillsides and valleys that later became the true centers of Clear Creek County silver mining. By the end of the summer of 1865, Frank J. Marshall and Nathaniel P. Hill staked claim to the Square and Compass lodes in the area later known as Silver Dale. Lorenzo Bowman and Jeremiah Lee also staked claims to parts of these lodes, the first known claims staked by African American miners in the Georgetown area. At the same time, those who were building roads and beating paths to the West Argentine area walked within feet of the valuable claims discovered around Brownville and Silver Plume over the next few years.[17]

By September 1865 the future of silver mining in Colorado shone bright. The riches of East and West Argentine, however, lay under several feet of snow for many months each year. The high-altitude location the early prospectors had sought was inaccessible for months on end and was not conducive to the needs of residential and commercial communities. As winter came to the high country, the hoards of prospectors returned to their homes in Central City, Idaho Springs, Denver, and elsewhere. Over the next twelve months, however, the valley around the old gold camp of Georgetown would take on a new life—"booming" in the best mining camp tradition.

The starting point became the flat ground at the foot of Leavenworth Mountain, also known as Burrell Hill. The dry ground and beaver swamps located between the two branches of South Clear Creek took on a new

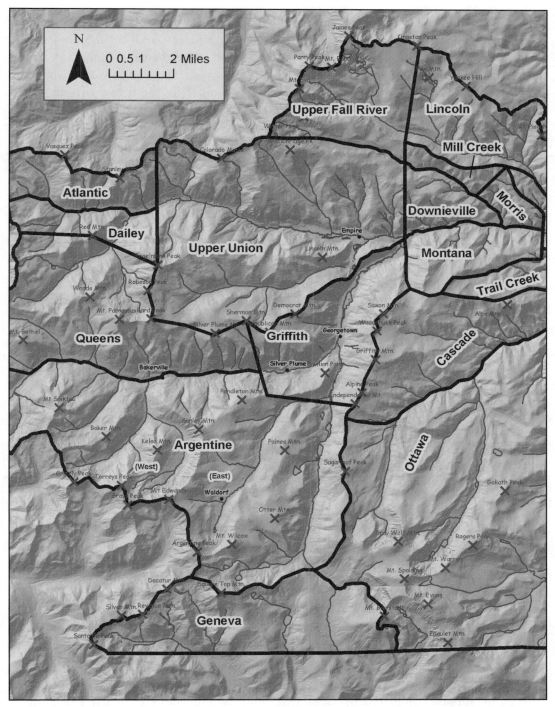

Map of mining districts in surrounding areas. Prepared by Matt Taylor, Mapping Department, Clear Creek County.

importance. Those traveling to the Argentine districts passed through the area, making it a logical location for homes, businesses, and mills. Supplies were shipped up to the mines, and ore came down, but the valley floor was largely unoccupied in the fall of 1865.

Ovando Hollister described the situation: "[In 1864] two citizens—Wm. Riley and W. M. Hale—and the few hands they employed, then constituted the population. . . . Still a year later than the time last mentioned, I was through here again, and although new life had been infused into Georgetown, it gave little sign of what it has since become."[18] Tax records from 1864 include only six individuals: Michael Doyle, John Dumont, George Griffith, John Harris, John Killion, and Stephen F. Nuckolls, along with two taxable structures or groups of structures belonging to the Georgetown Gold Mining Company and the Mount Alpine Company. The value of the town's "improvements" totaled $6,285, with "personalty" of $2,591—a small showing at best.

The silver craze of 1865 focused around Mount McClellan; however, the difficulty of transporting equipment and materials to the higher regions meant those who had buildings in and around Georgetown started to look at their future in a new, silvery light. Two significant events occurred in the fall of 1865. First, on October 26 William M. Hale, acting as an agent of the Mount Alpine Gold Mining Company of Colorado, preempted 140 acres of land

> being in the Township of Georgetown. . . . Commencing at the South East corner of the bridge next North East of the Mount Alpine Company's mill, thence in a line running Westerly to a stake marked two thousand (2000) feet, thence in a Southerly direction three thousand (3000) feet to a stake marked, thence Easterly two thousand (2000) feet to a stake marked, thence Northerly three thousand (3000) feet to the place of beginning, containing about one hundred and forty acres, being more particularly situated between the South and West branches of the South fork of South Clear Creek.[19]

Next, J. O. Stewart constructed a small furnace on the "left fork." "It was in the winter of 1865," wrote Stewart, "that Samuel Flint, Wm. Clifton and myself put up the first furnace in Georgetown, which was a small lead furnace . . . and from this little furnace we ran out 35 lbs of lead from several little lots of ore furnished by ourselves, Mr. Hepner, the Bells, Delamar and others."[20] The success of these ventures and the mines to the south and west furthered the growing interest in the community, but again the winter snows stopped development.

By the end of 1865 the Mount Alpine Company had invested thousands of dollars in the development of its mill and surrounding buildings. William Hale had the property he preempted surveyed and divided into lots and blocks. The survey, completed by George H. Hill, would define the early town site of Elizabethtown.[21] For many years, legend assumed that the Griffiths had named Elizabethtown in honor of a wife or sister; however, the Griffiths had left the valley by this time, and there were no wives, sisters, mothers, or grandmothers named Elizabeth.[22] In all probability, Hale named the fledgling town site in honor of *his* wife, Elizabeth.[23] The 1865 tax roll is the first known reference to Elizabethtown, shown as the location of the Mount Alpine Gold Company, with an assessed value of $28,950. Five new residences (John T. Harris, David Lees, J. E. Plummer, Robert Pooley, and B. F. Stephens) show up at this time, along with considerable holdings for the Georgetown Gold Mining Company and the Wilson and Cass Gold Mining Company.

Spring brought the start of a building boom in the mountain valley. If 1865 had been the "maddest, merriest" time for prospecting, the same could be said for 1866 and building in Elizabethtown. The *Rocky Mountain News* announced the new village:

> ELIZABETHTOWN.—This is the name of a new town that is springing up with great rapidity, a few hundred yards above Georgetown, on South Clear Creek, where the Argentine canon opens out abruptly, forming a beautiful park. Some fifteen or twenty new and neat frame houses are completed, and as many more are in process of erection. Many lodes are being extensively worked in that immediate vicinity.[24]

Hollister reported that fifty-seven dwellings, with an average value of $500, were built that year, along with three mills. J. O. Stewart's small furnace was joined by a more permanent business owned and managed by Caleb Stowell, and John T. Herrick and the Georgetown Smelting Company built a similar facility on the "right fork," later known as the Hall Mill. Joseph Watson, who had recently served as Central City's first mayor (elected in November 1864), turned his attention to the new area and started construction of a small mill near Chimney Rock that soon turned out a button of silver weighing twenty-three pounds, guaranteed to peak the interest of his friends and investors back in Philadelphia. The discovery of rich ore in the Anglo Saxon Mine, about two miles above Georgetown, started its own "rush" to the area with the news that valuable ore could be found close to town. The *Weekly Miners' Register* [Central City] voiced its support in September 1866: "It is now indisputably settled that the silver region extends all the way down to Georgetown. Assays indicate that the ores [there] are unusually rich."[25] The town was granted a post office, with William Gray as the first postmaster. During the fall, William Barton of Boston started construction of a two-story hotel, which must have towered above the valley floor as a sentinel of things to come. These new buildings, and the new faith in silver and its future, brought dreams for the future.

The developers of Elizabethtown generally came from the Northeast: Hale from Providence, Barton from Boston, Watson from Philadelphia. Most had served with the Union Army in the Civil War. They were generally well educated, well funded, and Republican, arriving in the West with an irrepressible enthusiasm. Their politics clashed with those of the early prospectors. The war was over, mining held the prospect of great fortune, and they were in a position

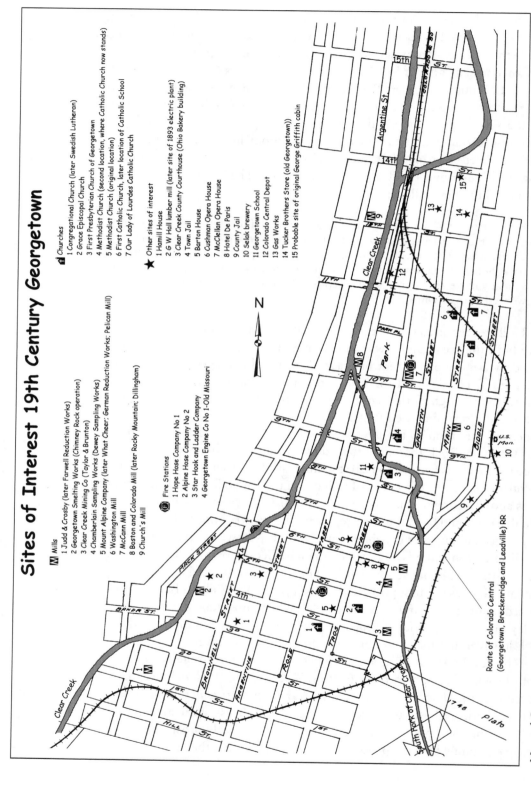

Sites of Interest 19th Century Georgetown

Mills
1 Judd & Crosby (later Farwell Reduction Works)
2 Georgetown Smelting Works (Chimney Rock operation)
3 Clear Creek Mining Co (Taylor & Brunton)
4 Chamberlain Sampling Works (Dewey Sampling Works)
5 Mount Alpine Company (later What Cheer; German Reduction Works: Pelican Mill)
6 Washington Mill
7 McCann Mill
8 Boston and Colorado Mill (later Rocky Mountain; Dillingham)
9 Church's Mill

Fire Stations
1 Hope Hose Company No 1
2 Alpine Hose Company No 2
3 Star Hook and Ladder Company
4 Georgetown Engine Co No 1–Old Missouri

Churches
1 Congregational Church (later Swedish Lutheran)
2 Grace Episcopal Church
3 First Presbyterian Church of Georgetown
4 Methodist Church (second location, where Catholic Church now stands)
5 Methodist Church (original location)
6 First Catholic Church, later location of Catholic School
7 Our Lady of Lourdes Catholic Church

Other sites of interest
1 Hamill House
2 G W Hall lumber mill (later site of 1893 electric plant)
3 Clear Creek County Courthouse (Ohio Bakery building)
4 Town Jail
5 Barton House
6 Cushman Opera House
7 McClellan Opera House
8 Hotel De Paris
9 County Jail
10 Selak brewery
11 Georgetown School
12 Colorado Central Depot
13 Gas Works
14 Tucker Brothers Store (old Georgetown)
15 Probable site of original George Griffith cabin

Route of Colorado Central
(Georgetown, Breckenridge and Leadville) RR

Map of Georgetown showing streets, with landmarks and mills. Prepared by Matt Taylor, Mapping Department, Clear Creek County.

to create a new, efficient, beautiful town nestled high in the Rockies. In all likelihood, many of these early settlers may have known each other prior to moving west.

Silver mining required the investment of substantial capital, and the East Coast had money to spare at the end of the Civil War. Much of that money came west with the sons of wealthy businessmen, who invested it in the new mines. Not all succeeded, but 1866 was a time of high hopes. Hollister provides some insight:

> Think of your correspondent riding out to this mountain cove [Georgetown] with the son of a man whose income amounts to $250,000 a year, whose name is in every mouth as an artist, a discoverer, a liberal giver to charitable and educational institutions—and that son a harder worker and poorer than I am! My driver also had to entertain me with the story of his loss of a fortune, and his "don't-care-a-damn-for-it," since that loss was the occasion of his striking Georgetown. He said, sage that he is, that a man never *lost* anything except his health. "A sound mind in a sound body," truly Heaven contains nothing better.[26]

The end of the war, the prospect of new technology, and an influx of newcomers with new money to invest, mixed with experienced businessmen from Central City and elsewhere, combined to create a new amalgam. The young silver mining community began to develop a personality of its own, one somewhat more genteel than those of the rowdier gold mining camps. The community quickly became very self-assured, focused on science and technology, racially and ethnically mixed, with everyone excited about the future.

The new breed wanted to become a permanent fixture and moved quickly to establish its position in the mining West. Within six months of the start of the building boom, with a population wavering in the hundreds and little permanent infrastructure, the newcomers gathered to move toward obtaining legal recognition. East Coast and overseas capital would be easier to secure if the village had a single name and legal recognition. The townspeople of both Georgetown and Elizabethtown knew they would have a better chance at success as a single entity. On November 24, 1866, the residents of Elizabethtown and Georgetown met to talk about their interests and the need to incorporate as a single town:

> Whereas, The growing prosperity and rapidly increasing population of this District, demand that some steps be taken to procure a legal incorporation or organization of the centres of population; and
>
> Whereas, It is impracticable to secure united action in measures of local interest, economy in public expenditures and uniformity in public improvements, without such action; and
>
> Whereas, The interests of the people of this district are one and the same, and union of feeling and action is the only means by which we can establish confidence abroad, in the value of our mines, and success of mining operations in our midst; and

Whereas, Every consideration of self interest conclusively establishes the fact that fair, just and impartial dealing should characterize the transactions of this community in local matters, it is hereby

RESOLVED, That the citizens of this District apply to the legislature for a town or village charter for a town or village organization, to be known as Georgetown.[27]

The name Georgetown had been in use for six years and would have been better recognized by the members of the territorial legislature. Incorporation was to be followed by application to the federal government for title to the townsite lands, which would require larger population figures than could be achieved by either town on its own. Besides, portions of Elizabethtown had been preempted from lands within the township of Georgetown, making it a subdivision in its earliest form.

The Mount Alpine Company preempted and platted the lands of Elizabethtown to secure its own interests—to generate revenue and to develop and sell lots itself. The company may have wanted to establish a new name that would not be associated with the failures of the Georgetown gold mines. Perhaps William M. Hale, a colonel in the Tenth Rhode Island Infantry, wanted to distance himself from the southern sympathizers in old Georgetown, an area later newspapermen would refer to as a "confederate cross-roads."[28] In any event, the townspeople focused on the need to develop "confidence abroad" and selected the name Georgetown over Elizabethtown for posterity. At the end of 1866 the town began to develop its formal presence. Committees were formed, assignments given. Whereas many early communities may have experienced a similar euphoria, Georgetown proceeded one step at a time, confident in its destiny and prepared for the work ahead.

No sooner had the mines been discovered and the town lots claimed than legal questions started to arise. The era of optimism soon ran into the reality of legal wrangling. It did not take long for the early discoverers to start bickering over claim lines while savvy businessmen from Denver and elsewhere elbowed their way into ownership of many of the best lodes and lots. In 1867 a letter to the *Weekly Miners' Register* bemoaned the pioneers' fate:

They have better legal advantages than we have—our lawyer not having arrived yet. They also have the advantage of advice from the land office officials. This enabled them to master the new national mining law [1866], and when they found how far they could go, they went their length—that is to say, they gobbled all the extra feet on all the best lodes in this district. . . . Thus by superior sagacity they have possessed themselves, at one dab, of as much good property as is now owned by the entire population of this district, all, too, at the trifling cost of a few half-dollars.[29]

Outside investors and experts, referred to as "soft-handed gentlemen," became a permanent fixture in the town and its mining industry.

Nevertheless, the new town and its new frame buildings impressed the same editors. In July 1867 the Central City newspaper reported:

Georgetown greets the eye and moves the imagination with the liveliest
pictures, as you reach the summit of the gentle elevation which hides it from
view until this point is reached. The buildings mostly new, and neatly
constructed, seem to have sprung up spontaneously from the earth like the
trees, grass and flowers.—None of that dingy, decayed, tumble down
appearance that so often marks the older mining towns, is seen anywhere
about the place. Bright cosy cottages are on either hand, painted snowy
white, and adorned with fresh green blinds and overshadowed with spruce
and pine.[30]

The "cosy cottages" sat within a growing business district. William Barton's
hotel perched atop a hillside in the center of town. Just to the south, George
Clark and Company, formerly of Central City, built a bank for the young
town. Across the street from the bank, Lemuel F. Yates built his house and
attorney's office. To the north of the Barton House, Tommy Barnes built a
saloon and pool hall. Alpine, Rose, and Taos streets became the center of the
business district, with Peck and Patterson's Store at the corner of Taos and
Alpine, Monte Bros. at the corner of Rose and Alpine, and the What Cheer
Mining Company—having purchased the property of the Mount Alpine
Company—at the head of Alpine Street.

The sons of New England's finest set out to create a town worthy of their
varied heritages. Although anxious to move to the new, open lands of the West,
these community builders brought their culture with them—quickly developing
the infrastructure and institutions familiar from their childhoods. As several
attorneys and surveyors worked on the final plat of the town, others formed
church congregations. The permanent vision of the community was greatly
enhanced with the first edition of the *Colorado Miner* in May 1867. Now
interested investors back in "America" could read about the promising future
of the new silver district.

Congregations of Baptists, Catholics, Congregationalists, Episcopalians,
Methodists, and Presbyterians quickly formed under the leadership of
missionaries heading to the Rocky Mountains in hopes of being the first to
secure the spiritual well-being (and building lots) of the new communities. Episcopal
Bishop George Randall's first published report summarized the situation:

> The ground here is *new* in every sense of the term. The Territory itself is but
> seven years old. The people have come hither from almost every state in the
> Union. They have severed old ties, and in many instances, former religious
> associations. Most of them have left behind not a little of prejudices, which
> bound them to certain systems, and made them hostile to others which they
> knew nothing about. They are now the subjects of new associations; are
> living a new kind of life; they are pursuing a new business, and have become
> so accustomed to novelties which they once opposed, chiefly because they
> did not understand them and did not care to understand them.[31]

Although the bishop stressed the newness of the territory and the numbers
of "undecided" members of the community willing to look at his church, it

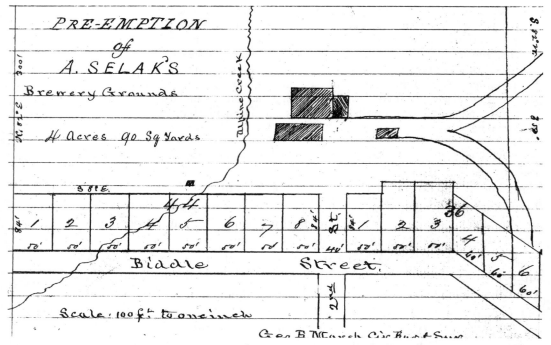

Preemption plat of Selak Brewery at the head of 2nd (now 9th) Street. Clear Creek County Archives, Book R, p. 406.

would be interesting to know if the bishop—who attended Brown University in Providence, Rhode Island—knew the Hale family or other New Englanders living in Georgetown at the time. New souls were the target, but the strongest supporters and builders of all the new congregations had been staunch members of their church communities before moving west.

In a similar vein, the proposed town government drew its basis from the Northeast, proposing a town board to be known as a *Board of Selectmen,* a term most commonly used in New Hampshire, Vermont, Massachusetts, Connecticut, and Rhode Island. Many such boards in the East operated without a mayor, selecting one of their members to act as chairman of the board, to run meetings, and to oversee the day-to-day functioning of the town government. The proposed Georgetown charter similarly ignored the position of mayor and assigned limited "executive" powers to a police judge, recommending that the first police judge be appointed by the Board of Selectmen and be required to be a justice of the peace. The police judge would run the board meetings, deal with the functioning of the town, and sit as a judge for town matters. The term *police judge* was used extensively in the early charters of mining towns in Colorado, generally interchangeably with the term *justice of the peace.*

Georgetown, however, appeared to be the only town that vested this authority along with the running of meetings and town affairs.[32]

By combining the traditional executive and judicial authorities, the balance of power was definitely skewed toward law and order. Technically, the individual who supervised the daily affairs of the marshal's office could direct an arrest, then preside over the trial. This "efficient," if somewhat less than impartial, system of law enforcement may have come from the early mining district courts in which the president of the district often served as judge of the court. Georgetown's system left little doubt that the townspeople would not tolerate the lawless elements that often plagued mining towns in the West.

After another meeting of the townspeople in the fall of 1867, the proposed town charter was formalized and forwarded to the territorial legislature. On January 28, 1868, the charter was accepted, and the Town of Georgetown was officially created by legislative action. The young village stood proud and ready to accept the mantle of authority and prominence it never doubted would come its way.

NOTES

1. *Georgetown Courier,* September 24, 1910, 5. In Robert Taylor's obituary the date of discovery is listed as "about the 15th of September." Ovando J. Hollister, in *The Mines of Colorado* (New York: Promontory, 1974, reprint of 1867 publication), 253, puts the date of discovery at September 14. The lode location filed at the courthouse says "on or about the 15th day of September." Book I, 501, Clear Creek County Records, County Archives, Georgetown, Colorado.

2. Many published reports of this prospecting party name "Robert Layton" as a member of the group instead of "Robert Taylor." The generally reliable contemporary author Ovando J. Hollister appears to have been the first to use the name Layton, yet all of the papers filed at the courthouse in October 1864 clearly reference Robert Taylor (including the original filing on the Gus Belmont, which clearly states the three locators as Huff, Taylor, and Steele). Anyone who has used handwritten records can see how "Layton" and "Taylor" may have been transposed. Nevertheless, all primary source material refers to Taylor. Layton's name does not show up in any of the grantor or grantee indexes or in any of the deeds. Two references in Georgetown newspapers confirm the name Taylor: the *Colorado Miner* (November 14, 1874, 3), reporting on an interview with Robert Steele and Jonathan Cox; and the *Georgetown Courier* (September 24, 1910, 5), in the obituary of Robert Taylor, which references his role as a member of the prospecting party and gives many details of the events of September and October 1864. Taylor does appear to have been the least known of the three men and does not seem to have remained in the Argentine area for any length of time, adding to the possibility for confusion.

3. Frank Hall, *History of the State of Colorado,* III (Chicago: Blakeley, 1889–1895): 319.

4. Upper Union District, Book B, 24, Clear Creek County Records, County Archives, Georgetown, Colorado.

5. Hollister, *Mines of Colorado,* 253.

6. *Colorado Miner,* November 14, 1874, 3.

7. The party included Robert Steele, Jonathan Cox, A. S. Patterson, Frank Dibben, James Huff, John Peck, and Robert Taylor. *Colorado Miner* [Georgetown], November 14, 1874, 3.

8. Book J, 236–237, Clear Creek County Records, County Archives, Georgetown, Colorado (document recorded October 5, 1864, at 2:00 P.M.).

9. Hollister, *Mines of Colorado,* 253.

10. Book F, 168–170, Clear Creek County Records, County Archives, Georgetown, Colorado.

11. Ibid., 170–171.

12. Ibid., 172.

13. *Colorado Miner* [Georgetown], November 14, 1874, 3.

14. *Colorado Miner* [Georgetown], January 12, 1878 (special edition), 4.

15. Bakerville and the area known as West Argentine became the center of the Queens Mining District.

16. See Erl H. Ellis and Carrie Scott Ellis, *The Saga of Upper Clear Creek* (Frederick, Colo.: Jende-Hagan, 1983), 54–77, for more information on the Baker Mine. This work provides a detailed, well-documented history of the area from Silver Plume to the Eisenhower Tunnel.

17. The settlement west of Silver Plume, referred to as both "Brownville" and "Brownsville" in oral history, will be referred to here as "Brownville," the spelling used in all official documents known to refer to the area—school district records, voting precincts, and contemporary newspaper articles.

18. Ovando J. Hollister, *The Silver Mines of Colorado, a Flying Trip* (Central City, Colo.: Collier and Hall, 1867), 4.

19. Book O, 421–422, Clear Creek County Records, County Archives, Georgetown, Colorado.

20. Quoted in the *Georgetown Courier,* February 10, 1906, 1.

21. The Hill survey is referenced in the record of early meetings between the residents of Georgetown and Elizabethtown; however, no known copy of the document exists. Reference to the original surveys can be found in the *Colorado Miner* [Georgetown], May 9, 1872, 1, as well as in contemporary copies of the *Black Hawk Mining Journal.*

22. *Georgetown Courier,* April 20, 1912, 5, contains two traditional folk stories about Georgetown that are incorrect: "The name Elizabethtown was given to upper town after Elizabeth Griffith, wife of George Griffith, for whom lower town was named. At a public meeting held in 1867, it was voted to merge the two towns under the name of Georgetown." George's wife was not named Elizabeth, and the meeting actually took place in November 1866, proceedings of which were formally published in the *Black Hawk Mining Journal,* December 1, 1866.

23. Reference to Elizabeth Hale was found in the 1850 census of Rhode Island, City of Providence, dwelling number 790, William M. Hale, Elizabeth B. Hale, and children. Civil War records were found on the Internet: "Civil War Soldiers and Sailors System Search Detail," M555 roll 3, indicating that William M. Hale was a member of the Tenth Rhode Island Infantry, entering service as a captain and mustering out as a colonel. This information verifies that he would be the individual referred to in the obituary in the *Daily Central City Register,* October 10, 1868, 4, as Col. Wm. M. Hale, formerly of Providence, Rhode Island, who died in Georgetown on October 8, 1868. The names of the cross streets in the new town reflected the influence of many early players: Hill Street (now First Street, after the surveyor and, appropriately, on the hillside of Leavenworth Mountain), South Street (now Second at the south end of town), Hale Street (now Third, after Colonel Hale), Burrell Street (now Fourth, after James Burrell, respected Central City businessman and president of the Griffith Mining District), Mary

Street (now Fifth, probably named after William and Elizabeth Hale's youngest daughter), Alpine Street (now Sixth, leading up to the door of the Mount Alpine Company Mill), and Grant Street (now Seventh, undoubtedly named after General Ulysses S. Grant).

24. *Rocky Mountain News*, August 27, 1866, 4.

25. *Weekly Miners' Register* [Central City], September 4, 1866, 2.

26. Hollister, *Silver Mines of Colorado*, 5.

27. *Colorado Miner* [Georgetown], May 9, 1872, 1.

28. *Georgetown Courier*, February 18, 1883, 3.

29. *Weekly Miners' Register* [Central City], February 19, 1867, 2.

30. *Weekly Miners' Register* [Central City], July 2, 1867, 3.

31. Quoted in Allen Du Pont Breck, *The Episcopal Church in Colorado 1860–1963* (Denver: Big Mountain, 1963), 35–36.

32. See, for example, the original town charters of Central City and Black Hawk.

Surveys

The early settlements of Georgetown and Elizabethtown had their own surveys: Georgetown's drawn first by David Griffith, then by Charles A. Hoyt, and Elizabethtown's drawn by George H. Hill. Definition of lots, blocks, and roads allowed individuals to know where to build houses while reserving rights-of-way for future roads. Communities that established their surveys early in their development had fewer problems with houses that later had to be moved out of the "street" (which wasn't a street before the town was platted).

Georgetown had two such problems: Michael Kelly's house, which stood at the east end of Third (later Tenth) Street, and the Markley building, which stood in the middle of Taos Street. The town fathers vacated the portion of Third Street in 1869 to allow the Kelly residence to stay where it was and authorized $600 from the town coffers to have the Markley building moved in July 1873 after several years of failed negotiations.

Although copies of these original surveys are no longer part of the town records, it would appear that the two townsites "met" at Eighth Street, where the right-of-way for Rose Street changed from 40 feet to 50 feet. The two surveys were very different. Old Elizabethtown's lots were uniformly 50 feet × 100 feet, with forty rights-of-way for the streets. Old Georgetown's building sites varied from long and narrow to square, with variable widths on the streets as well.

Elizabethtown's east-west streets were generally named after individuals, whereas Georgetown's streets were numbered. The town changed the entire system in 1882, dropping the names South, Hill, Hale, Burrell, Mary, Alpine, and Grant and changing them to First, Second, Third, Fourth, Fifth, Sixth, and Seventh. That same year Mack Street was renamed Spring Street, and new signs were placed around town.

George Marsh Jr. with survey equipment. Courtesy, Historic Georgetown, Inc.

Georgetown, as Others Saw It

Fifty-seven dwellings, averaging in value $500, were erected in Elizabethtown and Georgetown in 1866. These two towns, by late action of their inhabitants, are merged in one, and called Georgetown.

—Ovando J. Hollister,
The Mines of Colorado, 1867

The dozen or fifteen miles from Idaho, up by Fall River, Mill City and Empire to Georgetown, is quite the nicest bit of the inhabited portion of the mountains. Around and above this now thriving and most beautifully located of the principal mining villages of Colorado are many marvelously rich silver veins. It is one of the places that every tourist should visit.

—Samuel Bowles, *Our New West,* 1869

It reminded me slightly of a Swiss town. It is the only town I have seen in America to which the epithet picturesque could be applied.

—Isabella L. Bird,
A Lady's Life in the Rocky Mountains, 1873

Its appearance and surroundings are superior to those of any other mountain town. The town possesses an excellent system of water works, an effective fire department, five churches, an opera house and a graded school with some 360 scholars.

—Frank Fossett, *Colorado,* 1880

Looking north from the face of Leavenworth Mountain, ca. 1870. Courtesy, Colorado Historical Society, William G. Chamberlain photo, 95.200.1496.

Early view of the Georgetown Methodist church. Courtesy, Veronica Elliott, Elliott Family Collection.

Methodists

"The [Georgetown Methodist] church stood about three hundred feet from the base of the nearly perpendicular mountain, and the preacher's residence was about halfway between them.

"At that time cord-wood was bringing five dollars for a small load. The times were hard, for 'Burlah had not yet struck it.' We could not afford to buy at such prices. Fuel was needed at both places. There was plenty up yonder on the mountain in sight. How to get it down, was the question. Someone said, 'Let us make a wood-bee, and invite everybody to come.' It was accordingly done. A goodly number came with their axes, clambered up the steep mountain side, felled the trees, and sent them down the natural wood-slide, of a thousand feet or more. Before night there was wood enough to supply the church and parsonage for over two years."

—[REV.] I. H. BEARDSLEY
Echoes From Peak and Plain, 1898

Rise of the Silver Queen

GEORGETOWN'S FAME SPREAD WORLDWIDE IN THE mid-1860s. Over the next fifteen years the young village grew at breakneck speed in an effort to keep up with its reputation. Simple wood buildings were soon dwarfed by bigger, more elaborate structures. Brick buildings appeared in both the residential and commercial districts, which soon merged into one. Livery buildings were elbowed out of the center of town as the need for banks, dry goods, saloons, fraternal halls, and other businesses exploded.

Space and materials were at a premium. New residents brought the styles and ideals of their past. Immigrants from the New England states brought typical New England architecture. Houses, fences, and other details took on the appearance of what they had left behind—symbols of success, permanence, and propriety.

Georgetown, 1868. Joseph Watson's house (lower left, now known as Hamill House) stood as a symbol of mountain finery. The false fronts of the burgeoning commercial district are beginning to appear. Courtesy, Denver Public Library, Western History Collection, Arundel Hull photo, Z-5768.

Georgetown taken from Brownell Street, looking southeast, in 1871. The original Barton House has burned (January 1871), and the new structure has yet to be built. Courtesy, Colorado Historical Society, 10026580.

Above: *The first Barton House, 1868, built by William Barton of Boston in the fall of 1866, provided elegant accommodations for the town's visitors until it burned to the ground in January 1871. Courtesy, Denver Public Library, Western History Collection, Arundel Hull photo, Z-5766. Below: Mr. Barton quickly rebuilt after the fire, with his new hotel sporting a gambrel window. Courtesy, Denver Public Library, Western History Collection, Duhem Bros. photo, X-18670.*

This 1874 photograph was taken looking southwest toward Silver Plume. The white frame Methodist church is in the center of the photo. Across the street the one-story white building is most likely the first Catholic church and school building. The early Washington Mill stands at the photo's center left, fronting on Main Street. The roof is being installed on the town's new brick schoolhouse on Taos Street. Courtesy, Harold Frost Collection, Collier photo.

From the appearance of the streets, this Fireman's Tournament may have taken place in August 1877, celebrating the arrival of the train into Georgetown. The volunteer fire department kept the town from burning even though the commercial district was packed with wooden buildings. Note the photographer's head, lower right. Courtesy, Lee Behrens Collection, Chamberlain photo.

South side of Sixth Street, probably in the late 1870s. The early A. F. Curtis Hardware Store is easy to locate. The false-fronted building to the right, Spruance and Love (later Spruance and Hutchinson), stands on the location of the hardware section of Kneisel and Anderson's store. Courtesy, Denver Public Library, Western History Collection, McLean photo, X-1363.

Sixth Street, looking west, ca. 1870s, shows a rare view of the Delmonico Bakery building (later absorbed into the Hotel de Paris) as well as the front of the McClellan Opera House. The early commercial district seemed elegant to local residents of the time but has the image of the Wild West to the twenty-first–century eye. Courtesy, Denver Public Library, Western History Collection, Collier photo, gift of Leo Stambaugh, X-1442.

The view looking north from Leavenworth Mountain, 1874, with the new schoolhouse standing proudly on Taos Street and Rose Street opened to the north. Courtesy, Harold Frost Collection, Collier photo.

View to the southeast, late 1875 or early 1876, showing work in progress on Rose Street after the channel of South Clear Creek was straightened and moved to the west (out of the Rose Street right-of-way). Courtesy, Harold Frost Collection, Chamberlain photo.

Only a few years later (ca. 1878) the same photographer captured the same view showing more mills and new large homes with yards being built along Rose Street. Courtesy, Harold Frost Collection, Chamberlain photo.

The photographer aptly named this photo, taken ca. 1878, "Georgetown E of Rose St." The town has taken on an air of permanence, with the three-story Cushman brick building, stylish brick Catholic church, and residential buildings jammed along the streets. Courtesy, Harold Frost Collection, Weitfle photo.

"Georgetown, From the Gorge," a Chamberlain photograph ca. 1880, looking north across Chimney Rock toward the Catholic church. Courtesy, Harold Frost Collection.

Above: *Georgetown's role as a supply center for the surrounding mines and communities meant the streets were often filled with burros such as these. Photo ca. 1880 taken from Loop Drive looking south. Courtesy, Tutt Library, Colorado College, Colorado Room Photo Files PP 85-31s.* Below: *Business buildings filled the block next to the Hotel de Paris. This view shows the Ennis (formerly Leggett) House, Garbarino's Saloon, Girard (later American) House, and the west side of the McClellan Opera House. Courtesy, Denver Public Library, Western History Collection, X-1027.*

Mislabeled "Leadville, Colorado," this view of Taos Street (ca. 1880) shows the center of the lodging district with the American House on the left and Barton House atop the hill on the right. Courtesy, Harold Frost Collection.

View from the point of rocks looking over early Georgetown to the southwest. The false-fronted building along Main Street is probably the Tucker Brothers store. Courtesy, Denver Public Library, Western History Collection, X-1067.

Brick buildings in the mid-1870s signaled a new era for Georgetown's commercial district. William Cushman, local banker and investor, built both of these brick structures between 1872 and 1874, although the corner building was not completed until 1875 when Cushman added an additional thirty feet along the front and a third story over the entire building under the supervision of architect W.H.J. Nichols. Note also the location of the Courier offices (estab. 1877). The sign over the first brick building says "OK J D Lake OK." Courtesy, Historic Georgetown, Inc., donated by Cindy Sutherland.

3 "It Takes Two Years to Make a Mine"

MINING HAS A LURE, A "FEVER" AS IT IS OFT EXPRESSED, THAT PROPELS PEOPLE TO go places and do things they might never do otherwise. Samuel Clemens was one of those who, attracted by mining's lure, went west; subsequently, he became Mark Twain. Writing about his Nevada silver mining experiences, the would-be miner and millionaire described his "visions of silver" and reality.

> We climbed the mountainsides, and clambered among sagebrush, rocks, and snow till we were ready to drop with exhaustion but found no silver—nor yet any gold. Day after day we did this. These holes [prospects] were the beginnings of tunnels, and the purpose was to drive them hundreds of feet into the mountain, and some tap the hidden ledge where the silver was. Someday! It seemed far enough away, and very hopeless and dreary. Day after day we toiled, climbed and searched.[1]

That description could well depict the story of most fifty-niners and those who came after to the Clear Creek and Georgetown region.

The silver discoveries of 1864–1865 gave the district a claim to fame it had not enjoyed with placer gold and gave the young territory new mineral hopes. The failure of the gold speculation in 1863–1864 had hurt Colorado more than many admitted. Miraculously, a savior seemed to be at hand. Reality, however, soon overwhelmed Coloradans.

The problem was that silver was not normally found in its native state in streambeds or on mountainsides. Silver mining was hard-rock mining, digging into the ground following a lead, lode, or vein to strike the bonanza. Ultimately, this took knowledge, finances, and skill far beyond those required for placer mining. Timbering, drilling, blasting, draining, hoisting, and tunneling took experience. Skilled miners became a necessity. It also took money, time, and more than a bit of luck to turn a profit. Some disgusted individuals claimed more money had been put in the ground than was ever taken out in silver or gold.

The problems did not end at the mine portal. The ore trammed to the surface was complex. It took further knowledge and money to reduce it to a silver brick. Depending on the mine's location, ore had to be transported to a mill or smelter, which might or might not have the right process to reduce it. The odds were against the miner that his mine would become a bonanza.

The prospector who found a promising outcrop was just starting a time-consuming, costly process to make his fortune, which was why so many sold out. They did not have the resources; and for many the fun was in the search, not in the worry, work, and waiting involved in mining and reduction. Clear Creek County displayed this familiar pattern of miners leaving.

All this did not discourage those captured by the silver fever. They knew it was only a matter of time before they gained their fortune. The gamble, the excitement, the possibilities, the eternal hope drove them across valley floors and up mountainsides.

Providentially, as the men toiled up the steep mountains, they could dream about the old mining adage—the higher the silver strike, the richer the ore. The over 15,000-foot bonanza Potosi Mine in South America, which had been mined for centuries, gave the adage truth for the optimistic. Also, the veins were easier to see in weathered higher elevations, where the detritus of the ages had washed down the mountain.

The news that Colorado contained silver mines set off great excitement. The big news in western mining, once the 1859 excitement had calmed down, had been Nevada's Virginia City and its silver Comstock mines. Millions of dollars in silver had been dug out of these mines, and millionaires strolled along the town's streets. That silver output had helped finance the northern effort in the Civil War and produced wealth that dazzled investors and the public alike. The $50 million in production by war's end caught everyone's attention. Having given up his mining for the reporter's pen, Mark Twain thrived in Virginia City. He left a sparkling account of those days in his classic western story *Roughing It*. As Huck Finn said, "[H]e told the truth, mainly."

> Virginia had grown to be the "livest" town, for its age and population, that America had ever produced. The sidewalks swarmed with people. . . . Joy

sat on every countenance, and there was a glad, almost fierce, intensity in every eye, that told of the money-getting schemes that were seething in every brain and the high hope that held sway in every heart. Money was as plenty as dust.[2]

Nevertheless, the flush times eventually came to an end. Mark Twain left, and the Comstock seemingly faded in the mid-1860s.

Although the Comstock had faded, Nevada had gained statehood, and the railroad raced to reach it. The Comstock had sired prospecting, which had given birth to new districts throughout the state. Austin, Eureka, and a host of other districts beckoned alluringly.

Now it seemed Georgetown and its mines could become a second Comstock, joining a host of other western mining communities with the same aspirations. That was the hope Twain talked about and Georgetown dreamed about, which kept western mining moving for generations, from the deserts of Arizona to Alaska's Arctic Circle.

In their excitement these early prospectors and miners gave their "mines" (most were little more than prospect holes) names to match their expectations. Some reminded them of home, others of famous people or wealthy mines in other districts. A few named "their lodes" after themselves, a few reflected expectations of what they hoped to find, and some defied description.

The Omaha, Wheeling, New Boston, Wooster, Vermont, New Hampshire, and Providence probably indicate the places their owners once called home. Silver Glance, Queen of the West, Silver Point, Pay Rock, Silver Cloud, Hunkadora, and O.K. describe expectations. Famous people received their due—Henry Ward Beecher, Emmet, Calhoun, Tom Thumb, Pulaski, Pocahontas, and General Grant. Fewer mines were named after wives, sweethearts, or daughters—Dora, Alice, Elizabeth, Lily, and Cora Baxter. Brown, Baker, Griffith, and W. H. White were named after their discoverers. Some were named for already famous mines or were given a name common to mines—Phoenix, Calidonia, Seven-Thirty, Homestake, No Name, Muscovite, and Dives. Snowdrift, Choctaw, Dead Broke, Federal, Colorado Central, and Congress explain themselves. Garno, Huldah, Saco, Terrible, Big Indian, Nyanza, Accident, and Horrible leave the uninitiated wondering.[3] In any event, most mines lived only in the hopes of their developers and soon passed from the scene.

Georgetown went through the same high hopes, boom, reality, and problems experienced by nearly every mining district. Initially, hopes soared as high as the mountaintops. Georgetown did not have its own newspaper until May 1867, but Central City and Black Hawk newspapers served as willing promotional partners. Several enticing tidbits appeared in those papers during 1865–1866.

First, readers had to be convinced that not everything had been discovered in this promising district, "the most beautiful of any we have ever penetrated in the Rocky Mountains." Thus even as winter descended, in November 1865 the newspapers proclaimed: "Prospecting in Clear Creek County is still going on vigorously. Argentiferous galena is being found." The district contained abundant water power, fuel, ores that "smelt with ease," and plenty of room for "towns, mills, and even pleasure grounds." The richness and extent of the ores "have been abundantly proven." As a result, "[T]he silver mining of Clear Creek County is destined, we are convinced, to become one of the leading, if not *the* leading interest of the Territory." The article continued:

> Reports from the silver mines are excellent. Not a day passes but new discoveries are reported and extremely rich specimens are shown. . . . Work is being driven on something like a dozen lodes at Argentine. For silver the work of prospecting has but just begun. Thus far our mines will compare with any of the world.

Once some ore had been discovered and even the tiniest hole dug, promotion began. Announcements appeared of high-grade specimens "assaying from $50–1,000 a ton." Of course, mines had to be identified. The Baker Lode on Kelso Mountain was, proclaimed the *Daily Mining Journal,* "not only the best one in the district . . . but the best in Colorado" (quoted in *American Journal of Mining,* May 12, 1866, 1). The article continued that many were "fully believing it to be one of the great silver veins of the world. The Baker ledge is undoubtedly one of the wonders of the World in regard to quantity."

One report claimed "some 300 miners" were at work in the district in late spring of 1866, with assays running from $100 to $2,000 per ton in silver. How "extensive" the veins might be had not been determined, but "enough [had been] done to prove them workable." Even more promising news appeared the next year: "Every day or so we see the glistening silver coming over from the Georgetown district, which is enough to make one wish he had a pick and assay furnace, that he might let the pen scissors 'drop.'" Georgetown was, claimed an unnamed Colorado newspaper, "at present the center of silver excitement," with a "number of fine veins being opened" that promised "rich results."[4]

The hardy pioneers overcame every obstacle in their path. One report indicated, "Mining has begun, notwithstanding the rough weather, on the very crest of the ridge." Reliable old-timers heralded: "Altogether the prospects of the county are considered by old residents brighter than ever before."

Assay reports from Nathaniel Hill's Black Hawk smelter—the territory's most progressive and scientific smelter—were encouraging. They ran as high as $1,816 per ton on "selected ores." Ores from the Elijah Hise Lode assayed at $810 and those from the Adams Lode at $410, however, whereas some from the J. E. Lyon smelting works ran as low as $117. Although these amounts did not match the Comstock, they were still exciting for Colorado.

A reporter would visit a promising district and send back a glowing report—a "good deal of prospecting [was] now going on," lodes were paying "very well," and a "large, handsome mill" had been built. "No doubt" existed that the silver veins "are as rich as any in the world."[5]

A closer, more careful reading of the September 12, 1866, *Daily Mining Journal* suggested things might be less rosy than some portrayals indicated: "Nearly all the known lodes of the district [Griffith and Argentine] passed into the hands of Eastern men." That might have been good or bad depending on the investors' expectations, experience, and staying power. They organized companies to work their mines, then reality stepped in. Of the four companies started, the Mount Alpine Company could not make its mill work, and it reorganized. The Georgetown Company became "embarrassed" because it had failed; it sold out, and the property passed into other hands. Half of the new companies, therefore, failed within a year.

Freighting costs were high from these isolated, weather-controlled, lofty properties, and the mines had only been opened to a small depth—ten to fifteen feet in some cases. The owners soon learned that the ore near the surface was rich, thanks to the effect of water and weather leaching out base minerals, leaving behind precious metals. This "secondary enrichment" gave early promise, but then came the complex ore. Experiments were conducted on milling and smelting processes. Mountainous mining presented many challenges—elevation, transportation, weather, ore reduction, high costs, and questionable ore values compared with expenses.

A *Journal* reporter, writing on September 13, 1866, had reservations: "It will be observed that there is not sufficient exploration yet accomplished on the Argentine and Georgetown mines to very fully determine their character." He worried about the problems caused by absentee ownership: "If they are really anxious to do it [help Colorado through the pinched times] they must go to work soon."

Impatience and perhaps greed pushed the miners to expect too much too soon. An old Mexican proverb proclaims, "It takes two years to make a mine." The truth in that axiom appeared obvious unless one worked a mine rich from the "grass roots"; most mines took months or even years of hard work and investment to develop. That reality failed to satisfy eager Georgetowners or anyone else awaiting their bonanza.

Some mines seemed to have a promising future. Observant Central City newspaper reporter Ovando Hollister toured the district in September 1867, and his letters to the *Miners' Register* became the basis for the first book on Colorado silver mining, *The Silver Mines of Colorado, a Flying Trip*. Hollister faced the same problems everyone encountered in the mountains. He described one day's adventure: "I came down the 'dishonorable trail' from Argentine this afternoon to the imminent risk of my mule's and my own bones." He and his mule survived.

A few examples of what he saw will suffice, keeping in mind that he intended to promote the region as well as inform his readers. The Brown Silver Mining Company had opened three lodes on its property with an adit and a shaft. Its assays were rich in silver (the best at 370 ounces to the ton), plus lead that would help in smelting. Three thousand pounds of ore were packed out each day by jacks, at $16 a ton. The Anglo Saxon Lode on Griffith Mountain has "produced to date more ore and more money than any other in the district." With a main shaft seventy feet deep plus workings, it produced $12,000 in silver soon after it opened.

Most mines, though, were just "blast and pray" developments. The "Square and Compass" on Leavenworth Mountain had a shaft sunk "twenty-six feet" into a "kind of bastard quartz," in a "country badly shaken up." The Alice Jones on Griffith Mountain had "just been struck," with "excellent promise," Hollister wrote. The Astor, "on the summit of Democrat Mountain, a cold, bleak spot, where the timber has learned to crawl instead of climb under the incessant energy of the west winds," had a hole twenty feet deep. Over on Republican Mountain the Granada was "not improved enough," only a "prospect" hole on a thin vein.[6]

With the advantage of hindsight, in 1876 mining man Samuel Cushman reached a fairly accurate conclusion as to what had occurred in the 1860s. The excitement of 1864–1865 led to the "formation of several Eastern and home companies for mining and smelting." Although they were needed, the result had not been what was expected: "Some, with more faith and pluck than skill, attempted the impossible, and of course failed."[7] This reality reinforced the already preconceived notion of poor prospects for Colorado mining.

Hollister himself became somewhat disenchanted with the Georgetown District. In an article about Central City that appeared in the *Rocky Mountain News,* he included these comments:

> Speaking of Georgetown, I think I shall not go there, it is too much trouble. I went last fall, wore out my boots and my enthusiasm running over those terrible mountains examining prospect holes, wrote a description of the country and lodes then found in detail, which was first published in the *Register* and lost money by it. (*Rocky Mountain News,* November 6, 1868, 4)

Even so, he still predicted that times would be better with the arrival of the railroad.

Whether prospecting or mining, everyone had to keep an eye on the weather. Sudden storms in the mountains were a fact of life, rain bogged down transportation, lightning threatened anyone venturing about, and winter's snow slowed or stopped everything. Despite downplaying its impacts, locals could not ignore the weather. Comments such as this appeared: "[W]inter is settling in and work is slowing down," "many mills on Clear Creek are idle" on account of the weather, and "winter, or rather the fear of it, has put a stop to most [of] the surface work in the vicinity, and that at present comprises the greater portion done here."[8]

What could be done in such a situation, isolated high in the mountains without good transportation, short of developmental funds, and living under the dark cloud of the gold bubble? The tried and true answer, utilized by mining men and districts throughout the West—promote!

One could not depend on outsiders to do promotion, so the best answer was a local newspaper. Georgetown gained one when the *Colorado Miner* made its debut in May 1867. This newsy paper promoted, defended, explained, and reported on mining, mining interests, and local developments. Local miners and mines finally had the outlet they desperately needed in a West full of potential and promising districts that needed precisely what Georgetown needed to become rich.

The paper glowingly described rich ore and promising developments (mines and mills), adding that "times are improved wonderfully in Georgetown" and that "more confidence is now felt in the future of its mines than at any previous time." Coming events and developments sure to boost the local economy received their due: "We have it from good authority that reduction works on a large scale will be erected here soon." Items proclaiming that there "are very satisfactory reports from many prominent mines" on Leavenworth, Sherman, and Brown mountains became nearly weekly features.[9]

The *Miner* boosted the efforts of local miners when, for example, they hit upon the scheme of driving tunnels—which promised to lower the cost of mining, aid in drainage, uncover hidden veins, and ease the elevation problems. On the front page of the January 23, 1868, issue, the *Miner's* editor reported on the Georgetown Tunnel:

> We would like to see a dozen such enterprises in operation here, as our precipitous mountain sides offer superior inducement for works of this character. The ease of handling ores and draining mines which this mode offers should induce men of capital to invest in such enterprises.

A reporter for the *Rocky Mountain News* waxed excitedly about the tunnels: "The Georgetown tunnels are the keys which are destined to unlock the treasures of the mountains, and open a new era in the history of our Colorado silver mines."

The editor of the *Miner,* like his mining-town contemporaries, could make a disadvantage appear to be an advantage. The November 14, 1867, issue of the paper, for instance, commented: "Capitalists cannot find a more inviting field for profitable investment than our mining region now presents. Our miners are poor and properties can be bought here at rates varying from five to fifty thousand dollars." The poor pioneers did not have the funds and could not develop their properties, but they were willing to sell at a "rock-bottom" price to those with money and ambition. What an opportunity, what a bargain! The paper concluded, "[I]n a few years [these properties] will be worth . . . many millions besides being a constant source of profit to their owners." Concisely stated, that was the lure that kept prospectors searching, miners digging, and investors sending their money westward.

Letters to the editors of the outside press, such as the *American Journal of Mining,* also planted seeds of hoped-for local benefits. Continuing with the tunnel theme, this news item appeared in the October 31, 1868, issue of the *Journal*: "The great heights at which these Georgetown lodes are situated, will render tunneling necessary to their profitable development." Further, a tunnel driven in any "mountain about Georgetown will strike many lodes." All this created a "tunnel craze," whose results could be seen only in the future.

Sometimes a certified old-timer would write about what had occurred and about the town's glowing future. David Griffith of Georgetown wrote a long report to the *American Journal of Mining* on May 9, 1868. He had been a resident from 1858 to 1863, and he knew the town's past and present and offered strong ideas about the future: "As there are few men who can claim the title of 'Old Settler' in this Georgetown region with better grace than myself, permit me to speak particularly of the developments now being made." Griffith continued:

> Any of the main fissure veins [veins filled with minerals] of the country, whether small or showing more mineral wealth on the surface, they invariably increase in richness and quantity of vein matter as depth is attained. . . .
>
> I desire also [to say] that one of the most radical improvements that I have noticed in this country generally, is, that the mine owners have at last waked up to the truth that capital in the east is no longer to be drawn into a gold or silver speculation by the simple exhibition of a few pounds of auriferous ores, and acting upon this experience, they are going to work in earnest this spring to develop their own property, thereby showing their own confidence in their mines.

Having promised the reader that the mines would get richer and that the mine owners had learned their lesson from the 1863–1864 speculation, Griffith continued:

> They have also learned at last that if a mine has proved to be of real value, its owners will not have to go peddling after capital, when that capital can be represented at the mine. The result of this change in the *modus operandi* of mining operations here will be a development of the mineral wealth of the country during the coming summer, unknown to the experience of the past five years, and that, too, principally by home industry; for, in fact, the most important "capital" of any mining country possessing true merit is confidence in itself.[10]

"Confidence in itself"—Georgetown had that, and so did the *Colorado Miner*. It was only a matter of time, locals believed, until the district would rival the Comstock—hailed as the benchmark of American silver mining. Yet Griffith also pointed out that the district had needs. Among them were "dressing works" for the "picking, sorting, and washing" of ores so "they may be smelted with profit." He had, in fact, hit upon a "great need."

Georgetown, like its neighbor Central City, had run up against the problem that would determine its future—how to achieve economical reduction, with a high percentage of gold and silver saved. This had been the underlying problem that caused the gold bubble to burst, and Gilpin County had not resolved it. Easterners would only suffer through so much outgo, with little or no return, before they shut their pocketbooks.

The *Daily Mining Journal,* in an article entitled "Cheap Smelting for Silver," analyzed the possibilities Georgetown's miners faced. There had "been much talk of building furnaces in silver regions but all who have thus [far] tried have succeeded remarkably in not doing it." The reporter thought chlorides were the answer for galena ores but that for chlorides and sulphurets "amalgamation would doubtless be the preferable treatment." Something had to be done, or mining would cease: "Let us have their treatment, then, and our grand old hills for 300 miles in length and 100 miles in width will become as populous as bee-hives in early spring. What a magnificent field for development and new discovery!"[11]

Silver had been found, so only two matters needed to be resolved. Transportation had to improve, some of which could be done locally. The ultimate answer—the railroad—awaited outside capital and a Denver connection to the transcontinental routes. By the spring of 1867 a daily express wagon ran from Central City to Georgetown, carrying mail and a few passengers. That was not enough, and Wells Fargo planned daily stage service. The company's plans were stymied, though, because the existing bridges would not hold the coaches, and it threatened a lawsuit against the road builders to "have the bridges [made] safe." The threat worked. In August 1867 "hacks" arrived from Denver and Central City.

The second matter, even more important than transportation, was to develop an economical and successful reduction process that would free the silver from the gangue, the mineral associated with the ore in a vein. As discussed previously, every mining district, once the easily worked surface ores were gone, confronted this problem.

Gilpin County miners had run into "refractory" ores within a year of the district's opening. To their dismay, the simple crushing with stamps and amalgamating with quicksilver no longer worked. A great deal of money and effort was spent in the following years trying to solve the problem, and that, plus the stock speculation surrounding the gold bubble, had damaged Colorado's reputation. The answer would arrive with metallurgist and chemist Nathaniel Hill's Black Hawk smelter, the Boston and Colorado Smelting Company, which "blew in" in February 1868. That was in the future. Until then, Georgetown continued to struggle on its own.

Communication improved greatly when Georgetown "secured an extension of the telegraph line." The outside world was now only dots and dashes away.

In 1867 Central City's roaming newspaper man, Ovando Hollister, cataloged a litany of problems. The ore had been ground, amalgamated, lixiviated (leached by washing with a solvent), roasted, desulphurized, and nearly everything else anyone could think of—all without noticeable success. Hollister concluded, "A solution of the problem of treating our ores need not be looked for while every mine owner is endeavoring to adapt some method specially to his own ores."[12]

The miners remained perplexed about how to reduce the ore. Assay returns gave them a good indication of what they had, but when they took their ore to a mill, the return would be abysmally low. The difference in value rested in the tailings pile awaiting a better process. Miners blamed mill men, who shook their heads in dismay.

Near the surface, the miners had some success. Like gold, the silver ores were decomposed by nature. Unfortunately, as the miners went deeper the ores became as refractory as Gilpin County gold, but with a difference—the silver ores required even more complicated methods of separation.

In 1867 miners from Mexico tried their hand. Using an *arrastra* to crush the ore and a small adobe furnace to heat it, they attempted to copy what had worked in their homeland. That process, however, did not work in Georgetown.

Lorenzo M. Bowman, an African American from Missouri who had experience in lead mining and smelting, helped devise an initial solution based on furnaces. He developed a method that was successful on the surface ore, but the deeper ores—with which he had no experience—were more complex. Bowman also prospected on Leavenworth Mountain in 1865, discovering the Square and Compass lodes—some of the richest claims on the mountain—which he sold.

The *Rocky Mountain News* printed a short article about "Mr. Beauman, who had considerable experience in lead mining." He had joined some other "colored people of Denver" to organize a company to build a Scotch hearth and cupel furnace one and a half miles above Georgetown. The hearth and furnace would work the "valuable galena" ore from the lode they had purchased.[13]

A variation of that story appeared later. Bowman was credited with opening the Leavenworth District when he "headed a party of colored prospectors" into the area in the fall of 1864. They filed claims on a number of lodes, and Bowman "conceived the idea of building a smelter" and organized a company under "the name of the Red, White and Blue, composed of colored people only." The noted speaker, writer, and abolitionist Frederick Douglass became a stockholder. His son, Frederick Jr., served as smelter superintendent, with Bowman as "chief smelterer." Despite having plenty of ore, the experienced Bowman, and "the good will of the community at large," the enterprise "proved a failure, and the smelter was abandoned."

Bowman's involvement ended when he died in December 1870. His effort had been handicapped from the start by the lack of good fuel, skilled labor, and enough lead ores to keep his Scotch hearth smelter operating.

At least one other African American prospector actively worked in the district in 1868. W. H. White found a promising lode and, as was quaintly described, "bestowed his name upon the find." He sold his claim to an "association" that included Colorado's leading politician, Jerome Chaffee.[14] It was still being worked in 1876, although it was more promising than productive.

The Frieberg-educated (until the 1870s most trained mining engineers were products of European schools) and worldly William Bruckner arrived in Central City in 1867 to try out his ideas. Joined by other Germans in the area, he developed a plant based on roasting, a ball mill, and amalgamation. After achieving success with Georgetown silver ores, Bruckner moved to the community. Bad luck dogged his efforts, however. One mill burned down, and his process faced stiff competition from smelting ideas coming from the famous Comstock. His ideas later gained acceptance throughout the West, but they were premature in Georgetown.

The Georgetown Silver Smelting Company of New York presented another problem. The company was promoted by Central City's Jerome Chaffee, Eben Smith, and their partners, who built a smelter and displayed silver-lead bullion to help promote their mines. They made a good show and sold the property, but the plant failed to live up to expectations and quickly slipped into debt. In 1868 it was closed and sold by the sheriff.

Central City's *Miners' Register* had forecast back on October 2, 1866, that Georgetown would soon become a "city of smoke stacks and sulphur fumes." Happily, that did not occur; the only sulphur may have come from infuriated investors and mine owners.

The result of all this experimentation did not bode well for local mining, however. One disgruntled observer said the country between Georgetown and Idaho Springs looked like a graveyard of costly monuments: "These 'compliments to the dead' will not be found to resemble marble, nor nothing grand and imperishable, but rather will they be found after the pattern of ruined and ruinous mills, surrounded by old rusty machinery, and decked with scattered fortunes."[15]

Likewise, mining reporter Rossiter Raymond had little praise for the misguided efforts. In his 1870 government report he bluntly stated:

> The varying character of the ores necessarily involves the application of
> different processes to beneficiate each kind to the best advantage. All those
> ores containing much galena must be treated by smelting; those containing
> little galena with more or less zinc-blende, and the so-called surface ores,
> will be worked most economically by barrel amalgamation; and the poor
> sulpuret ores will afford abundant material for raw amalgamation. . . .
>   [Smelting attempts, with one exception,] have been metallurgical or
> financial failures. In some instances, this result was due to the management
> of the smelting operations by persons who seem to have believed that
> nothing was necessary to accomplish a success in that direction except
> common sense, fire-brick, ore, and wood. The results proved, of course, that

a thorough knowledge of chemistry and metallurgy, combined with practical skill and experience, are indispensable.[16]

That could have been the epitaph for a host of western mining districts in the 1860s.

The one exception, in Raymond's opinion, was the smelting works of the Brown Silver Mining Company. Under the management of Joseph Watson, the company initially made some wise decisions by hiring skilled men for the smelting, purchasing rich ore, and learning from previous failures. The company was just starting when Raymond made his enthusiastic report.

Unfortunately, it failed to work out well. Watson had a difficult time finding ore with enough lead to use in his furnace, causing the loss of large amounts of silver in the slag outside the smelter. Raymond had worried about this possibility even when he was optimistic about the firm's efforts. Also, the ore from the John Brown and other properties the company owned contained a high percentage of zinc, which interfered with the reduction. The result, as with so many early Colorado experiments, was that the mine failed within a little over a year.[17]

A recent scholar who evaluated Georgetown's pioneering smelting and milling efforts concluded that the early smelter men "had little knowledge of metallurgy and no practical experience." The reverberatory furnace many used was unsuited for local ores, which did not contain enough lead to make the process work successfully. The pioneers were further hampered by conditions in winter, which "was a difficult time for smelting ores in the mountains," here and elsewhere. Snow and freezing weather stopped or slowed transportation, froze water pipes, bedeviled machinery, and played havoc with mining overall.[18]

Easterners also criticized the experimentation and its many "processes." Yale professor and "geologist" Benjamin Silliman made pointed comments in a speech to the Connecticut Academy of Arts and Sciences in New Haven. Silliman, who supplemented his income by examining mines, had just returned from Colorado. He was not impressed, saying "Colorado is a mining country cursed by 'processes.' Mills are found in abundance filled with all manner of useless machinery, invented and constructed by men who knew nothing of the problem involved." He added that the stamps were badly managed and the methods of amalgamation " far poorer than in California."[19]

Another problem bedeviled the smelter and mill men: shortages of ore. In both 1868 and 1869, reports from Georgetown that appeared in the *American Journal of Mining* complained that mine owners had not supplied sufficient ore. The *Colorado Miner* blamed the problem on miners who were doing little to open their mines, explaining that some were doing "all in their power to develop mines, but where we have one such there are a dozen who will do nothing." The editor of the *Journal* concluded that these were "not very cheering statements truly."[20] The statements did show the frustrations and realities that had beset Georgetown mining.

Although the decision was not within their control, Georgetown residents and Coloradans in general wanted the territory to become a state. Statehood would, they believed, make them masters of their own destiny, with more political, promotional, and persuasive power. They believed Colorado was "rich in minerals and metals and ranges of pasturage—where 'cattle on a thousand hills' may feed for ages yet to come. . . . There is everything to hope for in her mining future."[21] Bypassed by the transcontinental railroad, the territory eagerly awaited connections. That development, combined with statehood, promised a future that gladdened everyone's hearts.

The railroad promised salvation. The iron horse would bring ease and speed to transportation, help lower the cost of living, entice investors to come, and assist in promoting Colorado, among other benefits.

As the decade neared its end, Georgetown's silver miners faced most of the same problems they had confronted four or five years earlier. Optimism had not died, however. As one reporter noted at the start of the spring 1867 mining season, "Georgetown appears to be the big card in the hill country. They are all rich there, but 'they can't see it.' Everybody is prospecting but nobody professes anything yet." Another writer pictured the scene, with "campfires, at night, dotting rugged peaks like a vast constellation of stars."[22]

Yet for all their trials and tribulations, these pioneers, "builders of empire," had transformed the wilderness and planted settlements. That, in their eyes, was a wonderful achievement. Hollister captured that accomplishment in true Victorian verbiage in his 1867 *Silver Mines of Colorado*:

> Now, in place of the scattered evergreens and great rocks that used to give such effect to the moon's white light, there are elegant residences; capacious and well regulated stores; tasty, roomy, convenient and well-kept hotels; banking houses and offices; reduction works whose wrangle never ceases; and no end to poorer structures, used as bachelors' halls or perhaps built only to hold lots. I confess to a good deal of surprise as Wells, Fargo & Co's coach rattled me through the closely built up streets [of Georgetown] where a year ago it was almost an unbroken solitude.[23]

NOTES

1. Mark Twain, *Roughing It* (Hartford, Conn.: American Publishing, 1872), 209.

2. Twain discussed in Grant H. Smith, *The History of the Comstock Lode* (Reno: Vanier Graphics, 1980 reprint), 58; ibid., 302–303.

3. The names were found in Ovando J. Hollister, *The Silver Mines of Colorado, a Flying Trip* (Central City, Colo.: Collier and Hall, 1867); Samuel Cushman, *The Mines of Clear Creek County, Colorado* (Denver: Times Steam, 1876); *Rocky Mountain News,* November 8, 1867, 3.

4. *American Journal of Mining,* "Colorado Mining News," May 12, 1866, 98; November 10, 1866, 98; March 30, 1867, 4; June 8, 1867, 206. The *Journal* was renamed the *Engineering and Mining Journal* and was one of the leading American mining journals in the late nineteenth century; *Rocky Mountain News,* November 6, 1868, 1.

5. The composite picture in the previous three paragraphs was taken from the *Daily Mining Journal* [Black Hawk], November 10, 1865, 3; March 3, 1866, 1; June 9, 1866, 1; September 12, 1866, 3; September 13, 1866, 1; October 4, 1866, 3; November 14, 1866, 3; November 26, 1866, 1.

6. Hollister, *Silver Mines of Colorado,* 10–11, 48, 50, 53, 54, 61, 67. That same year Hollister published *The Mines of Colorado* (Springfield, Mass.: Samuel Bowles, 1867), which contains little information about Clear Creek County but covers the whole territorial picture.

7. Cushman, *Mines of Clear Creek County,* 8–12. See also, Frank Fossett, *Colorado* (New York: C. G. Crawford, 1880), chapter 13.

8. *American Journal of Mining,* "Colorado Mining News," November 10, 1866, 98; February 23, 1867, 339; January 2, 1869, 4.

9. *Colorado Miner* [Georgetown], November 14, 1867, 3; January 23, 1868, 3; January 30, 1868, 1; August 27, 1868, 1; September 3, 1868, 1; *American Journal of Mining,* January 23, 52, and April 24, 1869, 261, quoting undated issues of the *Miner.*

10. *Rocky Mountain News,* November 8, 1869, 1; *Colorado Miner* [Georgetown], November 14, 1867, 3; *American Journal of Mining,* May 9, 1868, 29l; October 31, 1868, 276.

11. *Daily Mining Journal,* April 10, 1866, 1.

12. Hollister, *Silver Mines of Colorado,* 78–80. In the chapter on mining, definitions of mining terms were taken from Rossiter W. Raymond, *A Glossary of Mining and Metallurgical Terms* (Easton, Pa.: American Institute of Mining Engineers, 1881); *Rocky Mountain News,* May 14, 1867, 1; August 31, 1867, 1.

13. *Rocky Mountain News,* May 14, 1867, 1.

14. *Colorado Miner* [Georgetown], June 27, 1868, 1; August 3, 1878, 1; *Georgetown Courier,* February 28, 1914, 4; Cushman, *Mines of Clear Creek County,* 32–33; information on Bowman in *Rocky Mountain News,* August 2, 1867; July 6, 1890 (contains quoted material); Robert Spude, "The Ingenious Community," undated speech in author's possession.

15. *Rocky Mountain News,* March 26, May 14, 1867 (contains quotation); *Colorado Miner* [Georgetown], July 25, 1867, February 7, May 14, July 16, July 30, 1868; *Miners' Register* [Central City], May 28, June 4, 1867.

16. Rossiter W. Raymond, *Statistics of Mines and Mining* (Washington, D.C.: Government Printing Office, 1876), 374.

17. James E. Fell Jr., *Ores to Metals* (Lincoln: University of Nebraska Press, 1979), 60–61; ibid., 374–375.

18. Fell, *Ores to Metals,* 58–61.

19. *American Journal of Mining,* October 3, 1868, 282.

20. *American Journal of Mining,* May 9, 1868, 291; April 24, 1869, 261.

21. *Colorado Miner* [Georgetown], May 9, 1869, 1; *American Journal of Mining,* April 28, 1866, 72.

22. *Rocky Mountain News,* May 14, 1867, 1; May 23, 1867, 4.

23. Hollister, *Silver Mines of Colorado,* 48.

Lawyers

Lawyers were not beloved in mining communities. The old Nevada miner Mark Twain remarked that "to succeed in other trades, capacity must be shown; in the law, concealment of it will do." Many in Georgetown and elsewhere in the West concurred. Lawyers, in their estimation, seemed bent on depriving an honest man of the fruits of his hard work.

A letter to Central City's *Miners' Register* (February 19, 1867) left little unsaid. Why did Denver speculators have the advantage over the honest citizens of Georgetown? "They have better legal advantages than we have." Therefore "they gobbled [up] all the extra feet on all the best lodes of the district."

The writer continued: "We appreciate now more than ever since we were born, that we are a set of the most egregious asses in America. Here we have labored, and planned, and twisted, and gone on short allowance of grub for years that we might get hold of some property, while these silk-haired, soft handed gentlemen have got[ten] as much as we have, simply by a few strokes of the pen."

"The legal fraternity of Georgetown," posed in front of the law office of John H. McMurdy, probably near the corner of Fifth and Taos streets. Men identified from left to right: *Robert S. Morrison, L. F. Yates, Frank A. Pope, Frank DeLamar, J. H. McMurdy, Charles C. Post, Edward O. Wolcott, John A. Coulter, unknown man, and William R. Kennedy. Courtesy, Colorado Historical Society, gift of Mrs. Fred Holland, CHS-X4533.*

That outburst reflected a deeply held contempt toward attorneys in early Colorado. For example, the Independent District in Gilpin County forbade lawyers to take part in any case except those involving themselves. Other contemporary districts forbade them from owning property, and one went as far as to threaten to whip any lawyer found in the district.

Machine drilling in the Kelly Tunnel, Georgetown. Courtesy, Historic Georgetown, Inc.

Machine Drills

Mining, almost since time immemorial, has been physically hard, dangerous work. Miners at the time of the Roman Empire would not have felt out of place in the nineteenth century and could have adjusted quickly to the few changes made in mining practices since their time.

That situation would change after the mid-nineteenth century, first with the more general use of powder, then with the appearance of "giant powder," or dynamite. Even with these improvements, however, it still remained a case of man versus rock, particularly when hammering a drill into rock. As the Industrial Revolution roared into full swing, machines started to replace workers' labor-intensive efforts.

To drill with a single jack or to partner with another miner for double-jacking took effort, skill, and endurance. The answer was to develop a drilling machine, which Americans did in the 1840s. In post–Civil War America, power drills came into their own in the West, and Georgetown stood in the forefront of this development.

Initially operated by steam or compressed air, the drills proved faster and generally cheaper than hand drilling. Miners using a machine drill in the Baltimore Tunnel drove fifty feet per month, whereas a neighbor advanced eight feet by hand. Rossiter Raymond reported in his 1870 *Statistics of Mines and Mining* that drills "accomplish from three to four times as much as can be done by hand for the same cost, and in much less time." There was a tradeoff, however. Rock dust coated the miners' lungs, causing silicosis. Some of the skill and professionalism on which miners had prided themselves also disappeared.

Marshall Tunnel

It's dark as a dungeon, and damp as the dew,
Where the dangers are double, and the pleasures are few,
Where the rain never pours, and the sun never shines,
It's dark as a dungeon way down in a mine.
—MERLE TRAVIS, 1946

The dark dangers of the mine took a backseat on October 15, 1875, when the developers of the Colorado Central Mining Company opened the Marshall Tunnel to welcome the townspeople to a celebration acknowledging the opening of the Number 5 Lode. An estimated crowd of 800 to 1,000 wheeled their way in and out of the tunnel in ore carts that day to the huge cavern 500 feet below the surface and 800 feet from the mouth of the Marshall Tunnel. Stephen Decatur captured the moment for the *Miner:*

The waving light of hundreds of candles placed on either side of the floor of the tunnel, the bright glare of sunshine at its mouth, was a scene worthy [of] the pencil of the artist. And now music is heard for the first time deep down in the mine. . . . Away went stately dames and virgins fair with their lords of creation into the merry dance, and thus the festivities commemorative of a great event in mining affairs [were] fully inaugurated. An ample lunch with hot coffee and lemonade, was prepared by some one, of which everyone partook. No tangle-brain, granger juice, wine or beer was used on the occasion. . . .

We have met to commemorate a great event in mining affairs—the sinking of a shaft one thousand feet deep on this great silver mine, to explore its treasure vaults. Last July No. 5 Company was formed, and in two weeks the money, $50,000, was raised in Georgetown and Denver to commence this great undertaking. It is a Colorado enterprise. Capital will now understand that we believe in our own resources, and the effect will be that money will flow into the territory in sufficient quantities to develop our mines (*Colorado Miner,* October 16, 1875).

Miners at the mouth of the Marshall Tunnel with a well-dressed, unidentified couple. Potential investors, perhaps? Courtesy, Denver Public Library, Western History Collection, X-60935.

4 Birth of
the Silver Queen

When the coach containing generals Grant, Sherman and Sheridan rolled up
to the post office in this town, some diminutive souled individual burst out in a
spasm of groans. This is not the first time that rebels have had cause to groan
at the proximity of those generals, so they had no cause to be in the least
disconcerted.

—*Colorado Miner* [Georgetown], July 30, 1868, 4

WHEN THE GENERALS STEPPED OUT OF THE COACH, THEY SET FOOT UPON A YOUNG town poised for growth and excited about the future. The town's founding fathers had managed to incorporate the village in January 1868, then they cajoled the territorial legislature into scheduling an election the next fall to relocate the county seat from Idaho to Georgetown. The town's meteoric rise following the discovery of rich silver in the fall of 1864 drew residents from throughout the county and other parts of the state and nation. Suddenly, Georgetown's population overwhelmed the rest of the county, and the "powers that be" wanted a vote to bring the county seat to the new population center, especially since the residents of Idaho were getting ready to build a courthouse. An election was scheduled for September 1868.

On the local front, the town's first election in January 1868 put Democrat Frank Dibben (the assayer from Central City who had joined the early silver explorations of Robert Steele and friends in East Argentine) into the office of police judge, along with selectmen W. W. Ware (D) and Charles Whitmer (R) from the first ward (south end of town—formerly Elizabethtown), with H. K. Pierson (R) and John Scot (D) from the second ward (north end of town—old Georgetown). William Henderson (D) was elected the first marshal.[1] The board

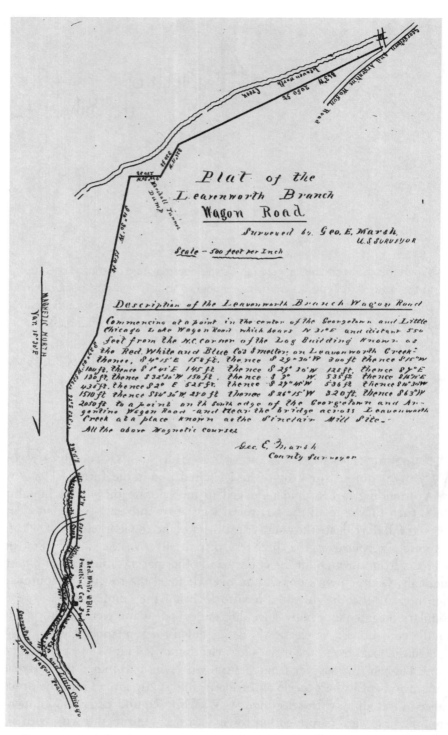

Plat of Leavenworth branch wagon road. Clear Creek County Archives, Book 21, p. 364.

met at the offices of the Baker Silver Mining Company (on Hale or Third Street between Argentine and Brownell) on February 17, quickly establishing four standing committees needed to govern the young municipality: Finance; Streets, Alleys and Bridges; Health, Police and Fire Department; and Licenses and Tippling Houses.[2] At the second meeting on March 13, the board instructed the Finance Committee to "take the necessary steps to purchase the Government title to the town site."[3]

Efforts turned to the upcoming election regarding relocation of the county seat, although the outcome seemed like a foregone conclusion if the legislature called for a new election, since the population of Georgetown was already twice that of the rest of the county combined. The election results bore out the new demographics: Georgetown 547, Mill City 90, Idaho 134, Grays Peak 1, Masonville 1.[4] Once the decision had been made, the county officers (several of whom lived in Georgetown) scrambled to find office space. Early plans for the erection of a courthouse were put on hold, however, as the voters in the rest of the county resented the young upstarts at the western end of the valley and convinced their friends and neighbors that things might change back soon. For the next six years, Georgetown held the title of county seat but was unable to have a courthouse until the Board of County Commissioners finally scraped together $1,050 to purchase the two-story wood frame Ohio Bakery building in 1874, which served as the county's home for the next 102 years.[5]

The game plan was complete. By the fall of 1868 the town had been formally incorporated, surveyed, and established as the county seat. Steps were under way to acquire the title to the land from the federal government. A newspaper, the *Colorado Miner,* was in place. Several congregations had been established, and three churches were about to be built. A burgeoning commercial district, a bank, and a fine hotel—suitable for dignitaries—greeted visitors. The townspeople had everything in place—everything erstwhile investors, future citizens, and "the right kind of people" would want and expect from the mountain community.

The local church congregations faced many dilemmas. First, each group had to find a place to meet that would be refined enough to encourage new members. Second, ministers had to be procured and convinced to stay. The Catholic, Episcopal, Presbyterian, and Methodist churches started by establishing early mountain missions that covered several towns, with a single minister as a "circuit rider." The Congregationalists soon followed, with new congregations springing up as soon as ministers could be convinced to travel west.[6] The task of filling these posts was far from easy, especially before construction of a railroad to Denver. A letter from Episcopal Bishop Joseph Talbot in 1863 summed up the difficulties:

> We could not but think how little conception our eastern brethren have of a journey or of the everyday incidents in a Missionary Bishop's life in the Northwest and wonder how one of the Bishops in our larger settled states

would look stretched upon a buffalo robe underneath his wagon fast asleep, and this in a public highway where the passing teams bestowed upon him a liberal share of dust. But necessity knows no law and even Episcopal dignity must give way before it.[7]

The bishop had started the trip with three associates but was the only one who made it all the way to Denver. Not only was the trip strenuous, but the lands were unknown, and many hesitated to bring families west until more was known about the territory. Many early missionaries returned to their homes in the East where they felt they could do more good preaching to hundreds rather than handfuls. Once again, Bishop Talbot provided insight after his appointment as bishop of Indiana in 1865: "Shall I work in a mission field containing, say, 50,000 souls? Or shall I take one with a population of a million and a third, and growing even more rapidly than the smaller?"[8]

The Methodists established the first formal congregation in Georgetown, with B. T. Vincent lecturing to the faithful at a meeting at Mrs. Plummer's boardinghouse in lower town on July 25, 1864.[9] Catholic Bishop Joseph Machebeuf gave the young mining town a vote of confidence in 1866 when he organized a mission in Georgetown, naming it "Our Lady of Lourdes." He occasionally ministered to the townspeople himself, as well as assigning Father Nicholas Matz and others to tend to the congregation.[10] By 1867 the Episcopal congregation had formally organized, and the Congregationalists were meeting on a part-time basis, with William Phipps preaching the services.[11] In March 1868 the Congregationalists formally organized, with five members, and commenced construction of a church that would cost $2,800 when completed.[12] The local Baptist congregation was formalized by Rev. T. T. Potter, an enthusiastic recruit who moved to Georgetown from Golden in the spring of 1868.[13]

The congregations scrambled to rent "appropriate" halls, changing to new locations as better buildings became available. The Baptists shared Reynolds Hall with the Methodists, a tiny building on Main Street in old Georgetown that had been moved from Empire. They later switched to McClellan's Hall on Alpine Street, along with the Congregationalists, when room became available. The Episcopalians first met in Patterson's Store at the corner of Alpine and Taos, then later moved, along with the Methodists, to Monti's Hall, a block to the west.

In August 1868 the Methodists announced plans to gather subscriptions to build a church on Main Street—located (not surprisingly) next door to the home of David Griffith, town founder and staunch Methodist who had recently returned to town and was a well-known lay preacher.[14] The building, built at a cost of $8,000, left the congregation with a heavy debt that was not paid off until 1882.[15]

By the spring of 1869 both the Methodist and Congregational churches were well under way, with the first service at the new Methodist church held on July 4, Rev. I. H. Beardsley of Ohio officiating.[16] Early stories indicate that some of the first calls to worship came by way of "unearthly shrieks on a conch shell" prior to the acquisition of a bell.[17] Although the Baptists never constructed a church, they did fascinate the local population by conducting baptisms in Clear Creek.[18]

Rev. Beardsley continued with the Methodist congregation until 1872, in stark contrast to both the Episcopal and Congregational flocks, which struggled to find permanent clergy. Rev. F. W. Winslow of Philadelphia held services for the Episcopalians in 1867 and the first part of 1868, with no regular replacement until the appointment of Rev. Courtlandt Whitehead, previously assigned to Black Hawk, in 1870.[19] The Congregationalists had a church building but no regular pastor until 1872. By September 1869 the struggling congregation was considering selling the building.[20]

The Congregational and Episcopal churches were located within a block of each other, set squarely in the middle of the New England transplants in the former area of Elizabethtown. Construction of the Episcopal church started in the summer of 1869 and, like the Methodist church, was within walking distance of many of its strongest supporters. D. H. Joy, a local resident, designed the church, which suffered a setback in November of that year when a Rocky Mountain zephyr knocked the half-built structure over against an adjacent building, delaying completion until 1870.[21]

In October 1869 Presbyterian missionary Sheldon Jackson arrived in Georgetown and called an organizational meeting that became the basis of the First Presbyterian Church of Georgetown. Although the church was started at least a year after the other five congregations, the group quickly secured the services of Rev. D. H. Mitchell and set about planning for the future. Early services were held at both the Congregational church and McClellan Hall.[22]

E. E. Tuthill became the first "long-term" Congregational minister, serving between 1872 and 1875. The congregation must have celebrated as the members watched him construct his lovely two-story home on Taos Street in 1873—a sure sign of his desire to stay. However, in January 1875 the *Rocky Mountain News* reported: "The Rev. Mr. Tuthill has preached his farewell sermon in Georgetown, and goes over among the heathen of California."[23] Rev. E. P. Wheeler assumed the duties, only to succumb to illness—leaving the congregation without a minister and with waning support from the American Home Missionary Society. No other minister was provided, and the congregation ceased to exist in 1882.

Amid all the positive growth and excitement in Georgetown, one event threatened the town's security. In late June 1868 a raging forest fire swept down the Leavenworth Fork of Clear Creek, turning onto the face of Leavenworth Mountain and leaving a blackened mess on the hillside. Backfires, ditches, and the dampening of roofs saved the buildings at the south end of town from destruction, but the threat was real and frightened many into packing up belongings in case they needed to escape.[24] Although reporters sounded calm in retrospect, the fear of trying to outrace the flames, fueled by potential winds whipping down the mountain valley, made an impact. The need for a fire department and proper equipment became paramount.

Protection from the threat of fire came in two stages. First, in March 1869 the town adopted ordinances that established guidelines for safe construction and operation of buildings: spark arresters on chimneys, double-wall construction, required space between buildings, and other items. The marshal, the police judge, and members of the board were given substantial authority to enforce these laws, as well as responsibilities relating to authority in case of fire.[25] Second, the first fire department was established in December 1869, accompanied by the announcement that the community had secured enough funds to order a fire engine, 300 feet of hose, and fifty fire buckets. The engine was ordered from Seneca Falls, New York. It featured a 7¼-inch cylinder, operating with fourteen men on the brake. The engine was guaranteed to throw a stream of water 130 feet, so the town fathers must have felt invincible compared to their earlier efforts, which had been limited to leather buckets. At last, the town's ready supply of water from the two creeks could be tapped in case of emergency.

The new piece of equipment piqued interest in the newly forming fire department. Interested parties were asked to contact "Captain" William H. Cushman at the bank of Jerome B. Chaffee and Co. (formerly the bank of George Clark and Co.).[26] In February 1870 a committee was appointed to draft bylaws and move for a permanent organization—just one month before the new pumper arrived. Members of the early department included H. C. Chapin, chairman; T. F. Simmons, secretary; William Cushman, treasurer; and committee members James Guard, W. W. Ware, J. Monti, J. A. Love, H. C. Chapin, J. F. Tucker, and C. W. Birdsall.[27]

By the summer of 1870 young Georgetown remained the center of activity, the "big city" in the county, with about 800 residents reported in the 1870 census. The *Miner* reviewed the rest of the county in May:

> Idaho has a population of 400 or 500 with four stores, two hotels, two bath-houses, where invalids and dirty folks may cleanse the outside, and two or three saloons, where the thirsty may soothe the inside. Fall River could muster a citizenship of fifty to seventy five. They have the Edwards House at Fall River, one of the best hotels in the Territory. Mill City is a village of possibly seventy-five inhabitants. There are two hotels and several stores. Empire City, four miles from Georgetown, has a population of one hundred and fifty. Good mines near there. One hotel, one store, and one or two mills. Bakerville is small. So is Brownville. Alvarado is maturing for birth.[28]

A "new kid" was about to show up in the neighborhood, however—a new town that would be closely tied to Georgetown, with shared interests in mining and tourism, throughout the rest of the nineteenth and twentieth centuries. In

Lithograph of Georgetown, 1874. Courtesy, Tutt Library, Colorado College, Colorado Room Photo Files PP 85-31s.

1869 Charles A. Kimberlin and Col. Ambrose H. Bartlett, of Doniphon County, Kansas, decided to develop a town—soon to be known as Silver Plume—in the small wide spot in the valley about half a mile east of the mines and buildings in Brownville. Things started slowly, with Kimberlin constructing only one or two buildings in 1869—although a contemporary reporter was impressed with the fact that he completed work on a school building before finishing his own residence.[29] Only a handful of buildings were built in 1870, with the Trenton Gold and Silver Mining Company, Kimberlin, and a Captain Cox leading the way. By the fall of 1870 there was talk of a pending hotel, and "[b]y consent of those interested, the place will hereafter be known by the name of Silver Plume."[30]

By 1872 the town was up and running, with a tightly packed commercial district, fifty or sixty residences, 400 to 500 residents, and plans for a town government.[31] The valuable silver mines on Republican, Sherman, and Brown mountains would produce great quantities of silver for the next forty or fifty years. Georgetown would become known as the "Silver Queen of the

Rockies," but most of the silver would be extracted from the mines above Silver Plume.

As might be expected, Georgetown's economic future started to shine with the location of major producing mines nearby. After several years of trying to figure out how to get ore down from the upper reaches of East Argentine, valuable ore was suddenly available just up the hill. Georgetown began to take on the appearance of a fine community. By 1872 brick buildings began to appear in the downtown. The young banker, William Cushman, started construction of two elegant brick buildings on Alpine Street. The town's early, rustic appearance soon gave way to fine homes, level sidewalks, a few trees, and elegant fences. The *Miner* described the progress:

> Internal improvements are all the rage in Georgetown. The streets and alleys are being patched up, many fences are being erected and painted, and door yards beautified, and we even notice that the dogs are being provided with new brass collars; all of which evinces a healthy degree of progress.[32]

The streets of Georgetown were packed with new arrivals from every point of the globe.[33] The Cornish miners, who had started to come to town in the 1860s, grew in number. Englishmen such as Robert Orchard Old, representing the British Colorado Mining Company, came to secure and protect their investments. The Garbarino and Sanguinette families came from Italy. James Gunn, James Owsley, and Frank DeLamar focused the concerns of the Irish by establishing a Georgetown chapter of the Fenian Brotherhood. Frenchman Louis DuPuy opened a restaurant in the Delmonico Bakery building.

The town included many African Americans, with special respect given to the early smelter of ore, Lorenzo Bowman—a veteran of the lead mines in Missouri—whose expertise helped start the early mining efforts in Georgetown. Bowman and Jeremiah Lee, a black man from Central City, started the Red, White, and Blue Mining Company south of town. Investors in the property included Clara Brown, a well-known former slave who had worked as a laundress in Central City to earn money to buy her relatives out of slavery. The network of black investors brought in money from a wide geographic base.

Black residents intermingled with whites on a regular basis in early Georgetown. Although segregation and discrimination certainly existed within the community, many comments from the local paper indicate a willingness of the races to work and meet together. Perhaps many white men from New England families, who moved out immediately after the end of the Civil War, had fought along with black battalions or been staunch supporters of emancipation. The *Miner* was incensed when local barber Charles Townsend was refused first-class service on a trip to Kansas:

> Our attention has been called to a letter written by C. O. Townsend, a worthy Colorado barber, a resident of this place, to one of his friends in relation to his treatment on the Kansas Pacific railway. It appears by Mr.

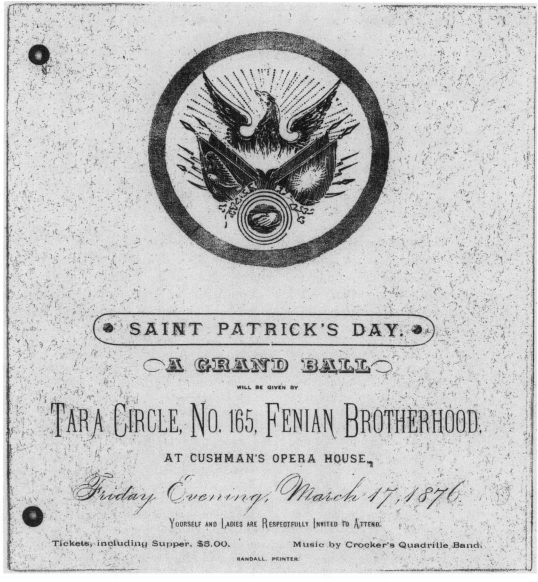

Fenian Brotherhood at Cushman's Opera House. Courtesy, Denver Public Library, Western History Collection, Randall Collection.

Townsend's letter that he purchased a first-class ticket in Denver, and besides being refused the privilege of eating his dinner at Salena, was compelled to accept second-class fare to Lawrence. If corporations sell first-class tickets the holders thereof are entitled to first-class fare.[34]

More insight comes from the description of the marriage of Mr. Townsend to Josephine Smith at the Methodist church:

Charlie Townsend has for several years been a thriving barber, and a respectable citizen of Georgetown. He was a soldier on the Union side, in the late war, went in when he was only 17 years old, and came out honorably. He is a graduate of Wilberforce College, Ohio. The bride is a niece of Aunt Clara [Brown], was educated at Oberlin, Ohio, and will make Charlie happy and not half try.

The colored people were out in force of course, but the white folks crowded the house. Father Cree gracefully got through his part of the programme, (except the kissing of the bride) and everything was done decently and in order. After the ceremony, the friends of the happy pair went to a restaurant on Alpine Street, which wasn't quite large enough; and from there they went to Charlie Yates old place and danced until daylight, and then went "home in the morning." Among the happy throng we noticed our gifted fellow citizen, Serg't George O'C-n-r and the agile artist John H. Oyster.[35]

Blacks owned property in town, including houses ranging from small, one-story homes—such as the cottage owned by Wade Townsend on Hill Street and the one of Gabriel Saunders on Ninth Street—to the more elegant two-story home of John and Mary Blackwood on Argentine Street. Annual celebrations included the anniversary of Emancipation Day (British West Indies). In 1873 several white residents joined in:

"The best laid schemes of mice and men gang aft aglee"—and it doesn't make any difference what the color of the men may be. The colored citizens of Georgetown, for instance, donned their best clothes, yesterday, and prepared for a grand celebration at the Devil's Gate. Clouds began to obscure the summer sky about noon, but the procession started, headed by the brass band. Down came the rain, and even emancipation ardor couldn't withstand the pitiless pelting of the sudden shower, and like a rope of sand the procession dissolved. A number afterwards assembled at the Silver Star Restaurant, where the occasion was celebrated in speeches by Rev. Tuthill, Prof. Weiser, Dr. Pollock and Mr. Olds. These gentlemen are all of the Caucasian race, and while they are all very good talkers, the entire absence of any African eloquence whatever made the occasion remind us very forcibly of the play of Othello with the dusky Moor left out. After the speaking, the orators of the day, and the members of the band, were invited to partake of an excellent dinner. The last we heard was the tuning of violins, from which we infer that the subsequent proceedings interested the toes more than the head.[36]

Through all the excitement and optimism surrounding the mining potential around Georgetown, a new, unrelated business appeared on the horizon—tourism. In January 1873 a headline in the *Miner* announced a new state industry: "At least one million dollars from Tourists in 1872."[37] From the earliest days of the town's development, visitors had come to explore the western frontier. Georgetown's merchants and hotel keepers readily sought the business of these adventurous travelers. One of the first attractions became the trip to Green Lake, an easy one-day ride up the narrow canyon that provided a pleasant combination of mild adventure and picnic opportunity. In 1867 the paper described the lake in glowing terms: "About two miles above Georgetown there is a beautiful lake, whose waters, clear as crystal when dipped up, shine with a tint of the most brilliant green in their natural basin."[38] The owners of the lake property soon purchased boats for rental and stocked the lake with trout. The construction of a wagon road to the property in 1872 made it more accessible to all:

> Green Lake is to be made a summer resort, a house is now being erected for the purpose of accommodating visitors, and will be opened in a few weeks. Strangers coming to Georgetown should by all means visit this beautiful and romantic mountain lake. The new wagon road is nearly completed, and when done will be quite a good mountain road.—There will also be a bath house with heating apparatus connected with the establishment. There are four boats on the lake for the delectation of the visitors. We hope the enterprising projectors of this public resort, Messrs. McCoy and Martin, will meet with success.[39]

Perhaps the most famous early visitor to the lake was Isabella Bird, the Englishwoman who rode up to the lake in 1873 and whose travels became famous with the publication of *A Lady's Life in the Rocky Mountains* in 1879. Her descriptions of the towns, mountains, and inhabitants of the West must have captivated readers throughout England. Her description of Georgetown as "the only town I have seen in America to which the epithet picturesque could be applied"[40] undoubtedly added the town to the itinerary of many of those headed to America.

The published travelogues of early visitors such as Louis Simonin, a Frenchman whose *Le grand-ouest des Etats Unis* described the West in 1867, stirred the curiosity of many in France. Publications such as *Harper's Weekly,* *Scribner's,* and others delighted in telling tales of rugged mountains, sleeping on buffalo skins, and other delights for those living in crowded cities. In 1878 Central City newspaper editor Frank Fossett published the first of three editions of *Colorado* (subtitled *Tourist's Guide to the Rocky Mountains*)—illustrating the growing market for visitors and adding greatly to local economies.

Scientists and surveyors numbered among the earliest visitors to the area. Botanists seemed anxious to venture up the high mountain slopes looking for unknown species of plants. Certainly, the lure of looking for plant specimens never seen or recorded by their contemporaries attracted the most adventurous of the trade. Both the mountains and the plants soon bore their names. In Clear Creek County, peaks were named for renowned botanists Asa Gray, James Torrey, Ellsworth Bethel, William Trelease, Edwin James, Charles Christopher Parry, and George Engelmann.[41]

As the towns and cities of the Rocky Mountain West continued to grow and prosper, the tightly packed streets became hazardous. A single spark or small fire could easily wipe out an entire commercial district. A single hand-pump engine could put out a fire in an individual building, but if the fire broke out in the tightly packed business district, the conflagration would outrace the efforts of any equipment that relied on water pumped from the creek. The lack of piped water in Colorado's mountain towns gave the flames a decided advantage.

Georgetown's first major fire and test of the new pumper came in Janaury 1871 when the Barton House, an elegant two-story wood frame hotel, caught fire in the middle of the night. The building was perched atop a hill at the corner of Taos and Burrell streets, and the sight of flames soaring into the air frightened the town's residents:

> The scene was an exciting one, neighboring buildings were beginning to scorch and smoke, sparks, and burning embers and shingles were picked up by the wind and conveyed to all parts of the town, the barn of the Barton House caught fire and for a time it looked as thouhg [sic] the town of Georgetown was soon to be in ashes. By the lurid light of the flames parties were to be seen removing their goods and household furniture from surrounding stores and dwellings and there was a good demand for wet blankets &c. to cover exposed roofs. Fortunately, however, for the safety of Georgetown, the wind lulled and died down in town, although the pines on the neighboring mountains could be heard moaning and creaking with the blasts that blew above us.[42]

The break in the wind and the fact that the Barton House stood alone on the hillside undoubtedly saved the town from further destruction. The new hand-pumper and young fire department performed bravely, but it became apparent that additional equipment would be needed to fight fires in two-story buildings. The townspeople quickly ordered hooks and ladders as well as a truck to transport them. The town fathers soon acquired property on the south side of Alpine Street, spanning Clear Creek, from William Cushman and Joseph Watson and started construction of the first fire building (prior to this time the equipment had been stored in rental property, also on Alpine Street).

In January 1873 the next major fire badly damaged the Stewart Silver Reducing Company about two mile north of town. A fire in Central City in May 1874 had even more impact, since it was instrumental in getting waterworks for Georgetown. Local businessman Albert Forbes later recalled that after hearing of the fire in Central, he closed his business and spent the afternoon figuring out how to organize a private water company. The June 13, 1874, edition of the *Miner* announced the formal incorporation of the Georgetown Water Company, with several of the town's most prominent businessmen on the board: A. R. Forbes, J. R. Hambel, Wm. M. Clark, Wm. H. Cushman, and C. R. Fish. As bankers, Cushman and Fish worked for firms that financed many of the businesses in town, giving them a direct interest in the development and operation of the waterworks.

The work started in July 1874, with a reservoir on the South Branch of South Clear Creek just above town and waterlines to surround the main commercial district, running from Hale (Third) Street to Alpine (Sixth). The town quickly installed seven fireplugs.[43] The original emphasis was on protection of the commercial, not the residential, district, since a fire in the tightly packed commercial area could destroy the entire town.

The installation of seven fireplugs in 1874 changed the nature and extent of fire protection in Georgetown: pressurized water could come from strategically placed hookups instead of firemen having to pump water from the creek. Thomas Guanella formed Hope Hose Company No. 1 on June 9, 1874. T. F. Simmons and the Alpine Hose Company No. 2 came onboard in November 1874, leading one early member of the Hope Hose Company to quip that it was one of the few times they had ever beaten the Alpines—and it didn't involve a race![44] The original company, Georgetown Fire Company No. 1, stayed in existence as an "engine company," concentrating its efforts on the town's pumper. Star Hook and Ladder Company had separated from Georgetown Fire Company No. 1 (also known as "Old Missouri") in February 1874. By the end of the year Georgetown had four separate companies, each with its own specialties: the hose companies attracted fast runners, the engine company required strength and endurance, and the hook and ladder company needed to anchor the ladders but also solicited young men who were not afraid of heights.

In 1875 the town acquired two additional lots and commenced construction of the Alpine Hose No. 2 on Mary (Fifth) Street and the "Old Missouri" Engine House further north along Taos Street. The engine house was strategically located—about halfway between the reduction companies on the north end of town and the residences that were starting to be built north along Rose and Taos streets. Both hose companies were located in the commercial district, where there was piped water. The Star Hook and Ladder Company stayed in the Hope Hose Company building for another ten years until its own building was built at the east end of Alpine Street, but it always remained in the commercial district, which had the tallest buildings and the greatest threat of devastating fire.

The arrival of the first hose carts excited the town even more than the completion of the waterworks. The first attempt to use the cart seemed like a holiday, as recounted in later years by H. H. Nash, a member of the original Hope Hose Company:

> In due time the new carts arrived, and we could scarcely restrain ourselves until the time set for making our first effort. But it came at last, and as I recall it, it was positively the most amusing thing I ever saw. We were out in force, and the streets were crowded. Any excitement in those days would bring out a thousand or more people in a few moments, and we had no reason to complain of our audience. We felt very proud, and were determined to do our best. No ten men would do for us in those days. No, sir. We had the whole company [between thirty and forty men] on the ropes to enable us to get there quicker. We were stationed on Rose street at the crossing of Mary, now 5th street; the whole 600 feet of hose on the reel. We

were to run to Alpine street, thence toward Taos street, unreeling 200 feet of hose and get water. We started and from that time the public thoroughfare would have been no safe place for the wildest of Texas steers or other wild animals. The street was not wide enough to accommodate anything other than our own team. We had selected a pair of wheelers, men of large avoirdupoise and No. 12 boots—designed especially to hold back rather than go ahead; as we figured that [was] a most important part of the work. We reached the plug in fair order and reasonable time, and then commenced the tug of war. Our wheelers threw themselves upon their haunches and tugged frantically in their efforts to stop the thing, while the men left the ropes and grabbed at the wheels, leaving the ropes to tie up the feet of the wheelers in a most dangerous fashion. But the process was successful finally, and the machine came to a stop. We then made attachment, and away they went again up Alpine street breaking coupling about Cushman's brick. Having placed the nozzle securely with a spanner, we turned on the water, and shortly had the satisfaction of seeing a magnificent stream playing over the highest buildings. I wish you could have heard the mighty cheer which arose from the throats of hundreds of citizens. Nothing was too good for us. Staid old businessmen like C. R. Fish, Uncle Billy Spruance, W. M. Clark, Joshua Monti, Silas Nott, Sam Strousse and others danced around like school boys, declaring no conflagration could ever get a start in the face of such work.[45]

Along with the "staid old businessmen" Nash mentioned, Georgetown's population included hundreds of young men—miners anxious to find their fortune or businessmen coming west to start out on their own, possibly learning a trade from friends or family. The fire departments provided camaraderie as well as protection. Fraternal organizations had formed within the town during the early years in an effort to provide friendship and support—Masons, Oddfellows, and Knights of Pythias, to name a few. However, some of the less traditional groups provide insight into what it must have been like for some of these young gentlemen to travel to the wild and woolly West, leaving their more traditional cultures behind.

How about the Celibacy Club, which met at young Henry Boyer's residence on Argentine? Or the Father Matthews Total Abstinence Club? The Sons of Temperance established a reading room, as did the Dickens Club, whose collections were reviewed by the *Miner*:

Among the monthlies we noticed Harper, Putnam, the Atlantic, Galaxy, the Overland, American Journal of Science, the Eclectic, Van Nostrand's Eclectic, Engineering Magazine, the North American Review, and the Manufacturer and Builder. Among the solid weeklies we noticed the San Francisco *Bulletin*, the Sacramento *Union*, the Memphis *Appeal*, New Orleans *Times*, Louisville *Courier Journal*, Chicago *Legal News*, Cincinnati *Times*, *Every Saturday*, *Harpers' Weekly*, the *Nation*, the *Engineering and Mining Journal*, and Forney's *Press*. Among the great dailies we noticed the New York *Herald*, Chicago *Times*, Chicago *Republican*, Missouri *Republican*, Boston *Advertiser*, and Pittsburgh *Commercial*. In concluding this brief and hurried notice, we desire to say that the young gentlemen of

Georgetown, who have so successfully organized the Dickens Club, have done a good action and instituted in our midst one of the great safeguards of society.[46]

By the mid-1870s Georgetown's residents had brought many of the "safeguards of society" into their everyday lives. The Presbyterian congregation completed construction of a solid stone church on Taos Street in 1873. The town constructed a beautiful two-story brick schoolhouse in 1874. The Catholic congregation finally outgrew the small 1872 wooden church and completed construction of a large, elegant brick church on Main Street in 1877. The Cushman Business Block included an opera house on the third floor. The town housed two newspapers and two banks. All that was missing was a railroad.

Talk of a rail connection to Golden, Denver, and the rest of the country started in the town's earliest days. Eager merchants knew they would have more tourists and more goods to sell if a railroad would come to town. Mine owners envisioned their ore traveling to smelters by rail instead of having to make the rugged trip by wagon. When the Colorado Central started construction of a narrow-gauge line up Clear Creek Canyon in 1871, with the intent of coming all the way to Georgetown, the town fathers had to be restrained from waiting at the end of town to look for the first puffs of smoke. The *Miner* shared the excitement:

> John Fillius informs us that the canyon of Clear Creek, west of Golden is alive with men and that three miles of the road bed of the western division of Colorado Central rail road is completed. We are permitted to say that it is the intention of the company to have the road bed completed to the forks of the creek by the 1st of November. There is a good time coming, boys.[47]

In April 1872 over 100 signatures were gathered on a petition to request an election to issue $200,000 in bonds for the Colorado Central Railroad Company (CCRR). The election was held on April 15, 1872, with 825 votes cast in favor of the bonds and 439 opposed.[48] The bonds were to be held by William H. Cushman of Georgetown, then paid to the Colorado Central in increments as the work on the line was completed according to a strict schedule. The company was unable to meet the deadlines, however, and the bonds were never issued.

The good times would have to wait another six years, as the railroad company struggled for funding and supplies and experienced several changes in management. The railroad did not arrive in Georgetown until August 1877. Throughout that time the *Miner* continued to sound positive, hoping to encourage the railroad and others interested in investing in the town. In February 1874 the paper reported: "We hear the question frequently asked: Will the CCRR be finished up to Georgetown the coming season? As a 'straw' we may mention that the editor of Crofutt's Western World learns from authentic sources that arrangemets have been made in New York City for pushing the CCRR through to completion with little delay."[49] In October 1876 another glimmer

McCLELLAN HALL.

CONCERT

WEDNESDAY EVENING, FEB. 3, 1875.

GRAND OPENING of the FAIR

—— FOR THE BENEFIT OF THE ——

CATHOLIC CHURCH.

As previously announced, the Fair and Festival will be opened by a Concert at which our best home talent will assist.

PROGRAMME.

"Viva Hibernia,"—*Gilson* ....By the Company.

"Five Hundred Thousand Devils,"—*Hoffman*,—Solo ....Mr. Kraatz.

"Mocking Bird,"—*Hoffman*,—Piano Solo.............................Mrs. Eldredge.

"Man the Life Boat,"—*Russell*,—Solo ....Mr. McNamara.

"Mynheer Von Dhunck,"—*Bishop*,—Glee, Messrs. Egan, Kraatz and McNamara.

"Kathleen Mavourneen,"—*Crouch*,—Solo ....Miss Phelps.

"Aurora Waltzes,"—*Chatterton*,—Duet ....Miss Cody and Mrs. Eldredge.

"Waiting,"—*Millard*,—Tenor Solo..Mr. Egan.

"Matrimonial Sweets,"—*Freeman*,—Duet......... Miss Cody and Mr. McNamara.

"The Harp that once through Tara's Halls,"—*Fowle*—Solo.............Miss Cody.

"How dear to me the hour,"—*Kleber*,—Duet .... Miss Phelps and Mr. Egan.

"Echoes from the Lake,"—*Violletta*,—Quartette,
 Miss Phelps, Messrs. Kraatz, Egan and McNamara.

ADMISSION, - - - - - 50 Cents

Audience are requested to assemble at 7½ o'clock. The entertainment will be opened at 8 o'clock precisely.

Catholic benefit concert at McClellan Hall, February 3, 1875. Courtesy, Denver Public Library, Western History Collection, Randall Collection.

of hope surfaced: "A party of eastern gentlemen, connected with the Chicago, Alton & St. Louis Railroad, visited the Silver Queen. . . . That this examination of our material resources may have a decided effect on the management of the Colorado Central, we devoutly hope."[50]

By March 1877 the headlines read: "THE RAILROAD COMING. The News Reliable This Time. Work Already Commenced. Three Cheers for President Loveland." The article went on to say that "one condition only [is] being stipulated for, that is, the people to secure the right of way through the country and furnish depot grounds. No money, bonds or relinquishment of taxes asked for—only co-operation on the part of the people."[51]

The town's businessmen immediately began collecting funds for construction of a brick depot building, to measure 50 × 100 feet. The anticipated cost would be between $3,000 and $5,000. In May 1877 the *Miner* had this update: "The finance committee reported the sum of $3,100 subscribed for the purchase of grounds. Hon. W. A. Hamill then guaranteed the further sum of $632.50, and the meeting raised a like amount. The committee should raise about $1,000 more, however, to make everything O.K., and we presume it will not be a difficult task."[52] Construction began immediately; completion of the line was scheduled for August 1, 1877.

The arrival of the train occasioned a grand parade and celebration, complete with races among the fire departments. A fine silver bowl and brass parade cannon were won by the quick-footed Alpine Hose Company No. 2, although the humor of the Star Hook and Ladder boys won the hearts of the crowd. The Stars knew they could not compete with the hose cart laddies' speed, so they put up their largest member: "'Jacky' came waddling down the track, the captain holding an umbrella over him and fanning him as he ran. When the score was reached, a couple of the Stars shouldered him and carried him up [the] street amid laughter and cheers."[53]

Anyone on the streets of Georgetown that day believed in the future of the young town. The fortunes of mining and tourism were theirs for the picking. Everything was in place; the town had everything it needed, and the future of the Silver Queen shone bright. In retrospect, however, the townspeople might have wondered if the train had finally arrived in town just in time to take people over the Continental Divide to the next biggest and brightest star on the horizon—the town of Leadville, whose silver production soon eclipsed the mines of the old Griffith District.

NOTES

1. *Colorado Miner* [Georgetown], January 30, 1868, 1.
2. Town of Georgetown, Proceedings of the Board of Selectmen, Minute Book no. 1, February 17, 1868, 3, Denver Public Library, Western History Department.
3. Ibid., March 13, 1868, 8.

4. Board of County Commissioners, Minutes, Book 1, Clear Creek County Archives, September 16, 1868, 238–240.

5. For more information see Christine Bradley, "The Life and Times of the Original County Courthouse," *Memberabilia* 1, 2 (August 1996), a publication of Historic Georgetown, Inc.

6. Walter S. Hopkins and Virginia Greene Millikin, *The Bible and the Gold Rush: A Century of Congregationalism in Colorado* (Denver: Big Mountain, 1962).

7. Quoted in Allen D. Breck, *The Episcopal Church in Colorado 1860–1963* (Denver: Big Mountain, 1963), 72.

8. Ibid., 28.

9. For an excellent review of early Methodist history in Georgetown, see the *Georgetown Courier,* July 19, 1913, 4, as well as J. Alton Templin, Allen D. Breck, and Martin Rist, eds., *The Methodist, Evangelical, and United Brethren Churches in the Rockies, 1850–1976* (Denver: Rocky Mountain Conference of the United Methodist Church, 1977), 214–215.

10. Thomas J. Noel, *Colorado Catholicism and the Archdiocese of Denver, 1857–1989* (Boulder: University Press of Colorado, 1989), 16. Central City was the first Catholic mission in Colorado, organized in 1861. Georgetown was the second mission. See also, *Georgetown Courier,* June 5, 1890, 3.

11. Hopkins and Millikin, *The Bible and the Gold Belt,* 40.

12. Ibid.

13. *Colorado Miner,* April 16, 1868, 4.

14. *Colorado Miner,* August 13, 1868, 4.

15. Templin, Breck, and Rist, *Methodist, Evangelical, and United Brethren Churches,* 69.

16. *Georgetown Courier,* July 19, 1913, 4.

17. Ibid.

18. *Colorado Miner,* March 31, 1871, 4.

19. *Colorado Miner,* June 16, 1870, 4.

20. *Colorado Miner,* September 30, 1869, 3.

21. *Colorado Miner,* June 16, 1870, 4.

22. *Steadfast in Faith: A Book of Memories, 1874–1974* (Georgetown, Colo.: First United Presbyterian Church, 1974), 10.

23. Quoted in Hopkins and Millikin, *The Bible and the Gold Belt,* 40.

24. *Colorado Miner,* July 2, 1868, 1.

25. Town of Georgetown, Proceedings of the Board of Selectmen, Minute Book no. 1, March 15, 1869, 82, Denver Public Library, Western History Department.

26. *Colorado Miner,* December 23, 1869, 4.

27. *Colorado Miner,* February 10, 1870, 4.

28. *Colorado Miner,* May 19, 1870, 4.

29. *Colorado Miner,* September 12, 1872, 1.

30. *Colorado Miner,* September 15, 1870, 4.

31. See *Colorado Miner,* December 7, 1871, 4; September 12, 1872, 1. Silver Plume incorporated in 1872, then disincorporated in 1874 (somewhat overwhelmed by the duties of municipal governance) before finally incorporating for good on August 19, 1880.

32. *Colorado Miner,* June 1, 1871, 4.

33. Information in this and the next paragraph from the U.S. Census Records, 1870 and 1880, and the Colorado State Census, 1883.

34. *Colorado Miner,* September 28, 1871, 4.

35. *Colorado Miner* (daily), February 28, 1873, 4.

36. *Colorado Miner* (daily), August 2, 1873, 3.

37. *Colorado Miner,* January 14, 1873, 2.

38. *Colorado Miner,* July 11, 1867, 4.

39. *Colorado Miner,* June 27, 1872, 4.

40. Isabella Bird, *A Lady's Life in the Rocky Mountains* (Norman: University of Oklahoma Press, 1960 [1879]), 190.

41. Erl H. Ellis and Carrie Scott Ellis, *The Saga of Upper Clear Creek* (Frederick, Colo.: Jende-Hagan, 1983), 28–30. Pages 33–36 contain accounts of Asa Gray's ascent of Grays Peak in August 1872.

42. *Colorado Miner,* January 12, 1871, 4.

43. *Colorado Miner,* July 4, 1874, 3.

44. *Georgetown Courier,* April 8, 1899, 3.

45. Quoted in ibid.

46. *Colorado Miner,* August 25, 1870, 4.

47. *Colorado Miner,* September 28, 1871, 4.

48. Board of County Commissioners, Minutes, Book 2, Clear Creek County Archives, April 10, 1872, May 22, 1872, 55, 64.

49. *Colorado Miner* (daily), February 23, 1874, 3.

50. *Colorado Miner,* October 21, 1876, 3.

51. *Colorado Miner,* March 31, 1877, 3.

52. *Colorado Miner,* May 26, 1877, 3.

53. *Colorado Miner,* August 18, 1877, 3.

Town Stats

INFORMATION WANTED.—Here is the sort of a correspondent we like to talk to—he affords us an opportunity to tell him something he doesn't know:

KNOXVILLE, Tenn., Sept. 9.

Editor Georgetown Miner: If not asking too much, will you send me a copy of your paper; also answer the following question[s]: 1. What is the number of inhabitants of your town? 2. What is about the price of land per acre around your town, and the price of town lots? 3. Have you any water power? 4. Is there much wheat raised around your town? J. B. Hoxie

1. We have in the neighborhood of 3000 inhabitants.

2. Land varies in price. Thousands of acres can be had for taking up, and then again many acres are valued at from $500,000 to $1,000,000 per acre. The first class fits mostly nearly on end, and is composed principally of rock not yet pulverized into soil, though some of it will do for mountain ranches and garden patches; the other class is occupied by rich mines. In addition to these, there are a few improved ranches, of limited extent, that can be bought at $15 to $25 per acre. Town lots range in price from $5 to $3000 according to location.

3. Water power is what we brag of, next to silver mines. It is almost unlimited.

4. Well, just in our immediate vicinity we do not raise much wheat. We don't brag on raising wheat in great quantities; but about 35 miles from here we can show friend Hoxie about the finest wheat growing country in the world. Georgetown is a mountain city, located in a small valley with granite hills rising from 1,200 to 1,300 feet above it on three sides; and our principal business is mining—the farming country lies east of the foot of the mountains, and can be secured at reasonable prices. We advise Mr. H. to come and look at it.

—*Colorado Miner* (daily), September 17, 1873

Sixth Street looking east, ca. 1875.
Courtesy, Colorado Historical
Society, Kirkland photo, 10030741.

Decatur

Stephen Decatur [Bross] penned many of the flowery phrases found in the early *Colorado Miner.* Decatur was nicknamed "Commodore" as well as "Old Sulphurets," and his associates enjoyed both his company and his flamboyant style. His love of the ladies is shown in an 1875 reference: "A sudden illness, coming on at an inopportune moment, prevented Sniktau [editor E.H.N. Patterson] from attending the Emerald Rifles' ball. 'O. S.' [Old Sulpherets] generously volunteered to act for us, with the following result—[he] fell in love as usual." His motto, according to one biographer, summarized his sentiments: "No orthodoxy, no monogamy, no monotony."

Decatur headed west after attending Williams College and teaching in New York for several years. No one knows why he dropped the prestigious last name of Bross (he was the brother of Illinois governor William Bross, among others), although most assumed it was to allow him to lead a less traditional lifestyle without embarrassing his family. Apparently, his former student D. H. Moffat, the Colorado banker and railroad mogul, recognized the wayward son, but Decatur refused to acknowledge his well-bred family.

Georgetown residents embraced Old Sulpherets as one of their own, graciously ignoring his multiple wives and checkered past. During his stay in town he constructed the Georgetown–Snake River Wagon Road, named the town of Silver Dale, and founded the town site of Silveropolis in Summit County. In later years Silveropolis was renamed Decatur in his honor.

Portrait of Stephen Decatur taken at Porter's Studio, Georgetown. Date unknown. Courtesy, Historic Georgetown, Inc.

Cushman

Few men in Georgetown promoted the young village more than William Cushman. His bank financed many of the early mining and milling ventures, investments that turned sour after economic downturns and the fires at the Stewart Silver Reducing Company in 1872 and 1875. Cushman built a brick building (now the home of Buckskin Leather) on Alpine Street in 1872, followed by the larger double-wide building at the northwest corner of Alpine and Taos (now the Clear Creek National Bank), which started as a two-story building, then built out to be three stories high. He was instrumental in the formation of the fire department and later worked to establish Green Lake as a resort worthy of attracting visitors from around the world.

In October 1871 Cushman bought out Jerome Chaffee's interest in the former bank of George T. Clark. In 1872 he completely reorganized and refinanced the bank as the First National Bank. One contemporary account indicated that Cushman and his father put $175,000 of their own capital into the venture. The new bank forwarded substantial capital to J. O. Stewart for his reduction works. The bank barely survived the first fire, then started to crumble after the second.

Cushman was arrested in New York City in 1880 after the funds held by the bank dissolved, causing many to lose all of their savings. The *Miner,* however, continued to speak well of the former banker, as seen in this excerpt from his obituary in 1886: "Mr. Cushman was a kind hearted man, and very liberal. He built some of Georgetown's best buildings, improved Green Lake and did many other acts that demonstrated his public spirit. During his bright days of prosperity, he was the most popular man in the city, and even after financial failure he numbered many warm friends who will regret to learn of his death."

William H. Cushman, ca. 1875.
Courtesy, Colorado Historical
Society, F-11791.

The national pasttime was popular in Georgetown, but the uniforms pictured here included ruffled edges that never caught on anywhere else. Two children, Will Shipley (left) and Jack Leyden (right), stand in front holding hands. Nellie Leyden, in striped dress, stands near Pat Leyden, to her right. Other team members (left to right) include Jimmy O'Donnelly, Jim Dinan, Billy Weavor, Fred Keeney, Henry Goetz, Frank Babcock, Jos Guanella, and Joe Trudeau. Courtesy, Denver Public Library, Western History Collection, X-1479.

Baseball

Baseball quickly became the sport of choice in Colorado's mountain communities (although in 1869 the "Snowy Range" cricket club of Georgetown competed with teams from other mining camps). In Georgetown blacks and whites competed on the same field, as members of the same team or as competitors: "The rival base ball clubs, the 'Sniches' (colored) and the Damfinoes (uncolored), had a match game Thursday, in which the 'colored troops fought nobly' as usual; but the white boys got away with their dusky rivals, by a score of 23 to 19. Our Professor— the 'Power of the Press'—was a little lame, or the result might have been different." The "professor" was Gabriel Saunders, a black man who worked for the *Miner* running the press.

—*Colorado Miner,* July 25, 1874

ONCE THE FORESTS WERE CLEARED, STREAMBEDS channeled, and wood cut for the winter, Georgetown's residents set out to create a life for themselves—complete with hunting, fishing, hiking, riding, and other types of play. The cool mountain air in the summer and the thrill of exploring a new and distant land brought visitors from afar (and not so far), but the local population enjoyed the "tourist" attractions as much as the guests.

Life in the Rockies

A hunting party returns to town with plenty of meat for the winter. Photo taken on Rose Street, looking north, ca. 1873, shows a new brick building on the left (west) side of the street. Courtesy, Tutt Library, Colorado College, Colorado Room Photo Files PP-85-31s.

The reason for camping at the corner of Ninth and Biddle streets is unknown—maybe the successful hunters simply couldn't wait for dinner. Courtesy, Denver Public Library, Western History Collection, Randall Collection, F-5269.

The beauty of the Colorado mountains has lured campers and hunters from the nineteenth century to the present day. Above: Campers with all the comforts of home: stove, bike, guitar, and picnic table. Courtesy, Historic Georgetown, Inc., Wheeler Collection. Below: Hunting party in Grand County. Courtesy, Historic Georgetown, Inc., George Dalgleish photo, gift of Betty Carman.

Crowds packed the streets (and the rooftops and balconies) whenever the fire laddies competed with rival towns. The two new Cushman brick buildings are seen on the left, but the building now known as the Cushman block is only two stories tall. An additional thirty feet was added to the front and a third story to the top in 1875. Courtesy, Tutt Library, Colorado College, McKenney photo, Colorado Room Photo Files PP 85-31s.

Fire department members posed proudly with their hook and ladder truck at the east end of Sixth Street, ca. 1875. Courtesy, Colorado Historical Society, X-1377.

Fire department racers pose on Sixth Street in front of the new brick section of the Hotel de Paris (ca. 1883). A note on one copy of this photo labels the second man from the left as William A. Hamill and says he used to pay the men ten dollars a day and put them up at Barton House, although they were under "strict supervision." Courtesy, Colorado Historical Society Library, 10026520.

Tourism became a major industry in Georgetown at an early date. Green Lake became one of Colorado Territory's best-known destinations. Above: courtesy, Denver Public Library, Western History Collection, Chamberlain photo ca. 1872, X-19353; below: courtesy, Denver Public Library, Western History Collection, Chamberlain photo ca. 1872, X-19298.

The quiet waters of Green Lake, three miles south of Georgetown, attracted tourists and locals for boat rides, picnics, fishing, and swinging. Left: courtesy, Historic Georgetown, Inc., Wheeler Collection; right: courtesy, Lee Behrens Collection.

Riding a burro to Green Lake was great sport and often involved a photo stop. Above: Unidentified riders pause for a photo in their best picnic attire. The town of Georgetown is visible in the upper left-hand corner. Below: Children ride with Fannie Tishler (left), Lena Strousse (fourth from left), Arthur Strousse, and Howard Strousse (right). The other two members of the party are unidentified. Courtesy, Denver Public Library, Western History Collection, Randall Collection, F-5339, F-5354.

Elizabeth Parker in front of Ermina Forbes's cottage on Fifth Street, behind Taos Square (early location of Forbes's Drug Store). Courtesy, Historic Georgetown, Inc., Wheeler Collection.

A jack party of women returning to Georgetown, probably from Green Lake, with lunch pails and babies. All the women are riding sidesaddle (no other style of horseback riding was acceptable for women at the time). Photo taken at the head of Third Street, ca. 1890. The discerning eye will find young boys poised on the rock cliffs behind the riders. Courtesy, Historic Georgetown, Inc., Wheeler Collection.

Nineteenth-century tourists and wagon masters were the predecessors of today's four-wheel drive enthusiasts without the motor. Above: Fully loaded wagon heading north toward Georgetown on the Summit County side of Argentine Pass with Grays, Torreys, and Kelso mountains in the distance. Courtesy, Denver Public Library, Western History Collection, Collier photo, C-162. Below: A horse trip to the summit of Grays and Torreys mountains inspires a photo. Courtesy, Denver Public Library, Western History Collection, Stiffler photo (1884–1892), Z-2549.

Entitled "Gray's Peak Number Four," this shot shows travelers posed on a rock outcropping near the summit of Grays Peak with the ridgeline of Mount McClellan in the background. Courtesy, Denver Public Library, Western History Collection, Stiffler photo, Z-2554.

Left: "Norman L. Patterson, Drum Major, Co. C, 1st Infantry C[olorado] N[ational] G[uard]." Courtesy, Denver Public Library, Western History Collection, Randall Collection, unnumbered. Above: The Georgetown Glee Club (looking anything but gleeful), September 8, 1890, with only four names listed: P. Ashcroft, Geo. Rubado, J. Keck, and T. Rodda. Courtesy, Denver Public Library, Western History Collection, Randall Collection, X-6537.

Above: *A guitar and accordian complete the ensemble. Courtesy, Denver Public Library, Western History Collection, Randall Collection, F-5203.* Right: *Portrait of young Harry Nash, later a longtime member of the Hope Hose Company. Courtesy, Historic Georgetown, Inc., Wheeler Collection, Porter's, Georgetown, Colorado, photo.*

Above: *Like a scene from* The Music Man, *young men from the town pose in front of the Georgetown School with their instruments. The woodwind section may have been difficult to hear, as the ensemble included only one clarinet, two drums, and the rest a combination of cornets, trombones, and other brass instruments. Courtesy, Denver Public Library, Western History Collection, X-1422. Below: A class picture, taken outside the Georgetown School, with children of all ages. Two young African American students are part of the group. Courtesy, Denver Public Library, Western History Collection, Randall Collection, X-1375.*

5 "The Mine Is Yet in Its Infancy"

GEORGETOWN'S MINES HAD PROMISED MUCH BUT DELIVERED LITTLE AS THE 1860s came to a close. Too many problems—isolation, lack of reduction facilities and profitable methods to handle low-grade ores, financial shortages, poor transportation—negated the fact that gold and silver had been found. To complicate matters further, the gold bubble disaster had made Colorado a pariah among eastern investors. Profits from the Civil War created millionaires and enriched other northerners from all walks of life. Colorado's reputation did not entice them to venture there, and local mining languished in the backwater of despair.

Why would eastern and foreign investors put money into Colorado mining when so many opportunities were closer and more promising? Railroads, iron and steel mills, oil wells (wells in Titusville, Pennsylvania, had produced fortunes during the war), construction, real estate, coal, copper, and iron mines all beckoned closer to home. For the risk takers, southern cotton and northern grain speculation offered a wild ride.

The depressed Colorado was not alone among western mining states and territories. The bright star of the early 1860s—Virginia City and its Comstock bonanza—had hit borrasca and with it the rest of Nevada. Montana had seemed promising, then its gold placers declined, and the hard-rock mines suffered from the same troubles as Georgetown. Isolated Idaho could not pick up the slack; its placer deposits followed the usual pattern after the first boom production dropped steadily. Once the poor man's bonanza, California mining had become industrialized with the now old forty-niners squeezing out a living as relics of another age. The Apaches kept Arizona mining to a minimum, and Mormon Utah had attracted little attention. Wyoming's South Pass excitement barely counted.

Optimism about the present and the future seldom waned. A correspondent to the *Rocky Mountain News* confidently told readers, "Each visit of the writer to the silver centre reveals to him progress in the great work of opening up the great riches which lay imbedded in the lofty mountains."[1]

For the eternally hopeful, however, help seemed on the way, and in truth it did come closer. The transcontinental railroad was being built from both the east and the west, and the Union Pacific and Central Pacific met at Promontory, Utah, on May 10, 1869. Shockingly from Coloradans' perspective, the railroad bypassed the territory, but Denverites were building the Denver Pacific to tie into the line near Cheyenne, Wyoming. It appeared that rails would soon crawl into the mountains. That would end the isolation, and blessings would shower over Georgetown. As the world rushed in, would Colorado statehood be far behind?

Nathaniel Hill's Boston and Colorado Smelter was successfully up and running in Black Hawk, not far from Georgetown as the crow flies but too far for low-grade ore to be transported over poor, weather-dominated roads. On a positive note, Hill tackled the gold and silver reduction problems from a scientific, not a school of hard knocks, approach.

Better transportation would, in theory, lower the cost of living and by that also lower wages, provide economical shipment of ores, bring investors to the mining districts more easily, and transport the latest industrial methods and equipment to Georgetown's doorstep. The optimistic believed a boom was at hand.

Nevertheless, the railroad had not yet even reached Denver, and as 1868 dawned it languished many miles and even more dollars away from the desperately needed connection to the transcontinental line. It would take several years, at the least, to get the rails to Clear Creek County after that connection had been made.

Colorado also needed to prove to the world that it actually had rich mines in various locales. Skeptics, even those with short memories, recalled the failed promises of 1863–1864. At least one district or, better yet, several districts not connected to or tainted by the gold bubble fiasco had to be opened. Having Clear Creek, Summit, Gilpin, Boulder, and the other older counties try to reinvigorate old mines would not suffice.

With a new, promising district, perhaps investors would flock to the area again, and Colorado would regain some of the mining momentum lost as the placers declined and the stock speculation burst. Georgetown's mining awaited the dawn of better times—times it could not create alone.

Georgetown residents and their mines were not rugged individualists who could stand alone. They were not masters of their destiny. The future rested on the efforts of the prospectors, miners, and mine owners.

Robert Old, an influential Georgetown mining man, was one of those determined to promote his town and Colorado. Disgusted with the gross inefficiency, mismanagement, overcapitalization, and excessive speculation that had characterized much of the 1860s, in the fall of 1868 he organized the British and Colorado Mining Bureau. His English office enticed investors with ore specimens, mining prospectuses, Colorado newspapers, and two pamphlets Old wrote entitled "Colorado"—with slightly different subtitles—published in 1869 and 1872.

Before it closed in 1873, Old's bureau helped produce some mine sales to foreign capitalists, provided a clearinghouse for the most reliable information that could be found, and conducted a "spirited organized publicity campaign." To him "go[es] the credit for the first successful systematized [effort] to interest foreign capital in the mines of the Rocky Mountain region."

Old became superintendent of the Terrible Mine and owned several mines that produced fairly well. He continued to write and speak about Georgetown in the years ahead, a "self-appointed ambassador." One of the wealthiest mining men who stayed in Georgetown, Old was president of the school board, an ardent Mason, and a charter member of the Pioneer Association of Clear Creek County. He was actively involved in politics, and the Free Silver issue (an effort to raise the price of silver) later became his consuming passion.

Meanwhile, as 1868 dawned, by conservative estimates Clear Creek County mines had produced approximately $2.14 million in gold and silver since the Pike's Peak rush. During those years the price of gold was set internationally at $20 and change per ounce. The price of an ounce of silver hovered around $1.33, floating on the world market. As an indication of the times, silver profits in Clear Creek County over the last two years had averaged only about $70,000. In comparison, during that period Colorado's number-one mining county, Gilpin, had a total estimated production of $2.7 million. For a new mining territory these were reasonable, if not spectacular, figures. In contrast, the Comstock's Ophir Mine alone had produced $5.294 million in silver during the 1859–1867 period.[2]

The chance of finding a mine equal to the Ophir or any of the Comstock's major producers lured prospectors and miners to explore Republican, Griffith, and Leavenworth mountains. No matter how remote the possibility, people kept searching throughout the West.

Government reports helped fuel such speculation. In 1868 James Taylor, for example, wrote excitedly about Clear Creek County's potential:

The prevailing great richness in silver in the ores of Griffith and Argentine districts . . . first attracted particular attention to the element of silver. In these districts silver ores of great richness have been discovered, masses being exhibited at the Paris Exposition, from the Baker Lode of Argentine District, and of the Elijah Hise and Endigo lodes of Griffith District, which assay respectively, in silver alone $532.12, $1,656.20, and $1,804.83 to the ton of 2,000 pounds of ore.

Taylor also praised Summit County and the "good" roads that were being constructed.[3]

Two years later Commissioner of Mining Statistics Rossiter Raymond went into more detail about Georgetown. Near the surface, silver ore was "more or less decomposed," where weather and water had flushed out some of the minerals that hampered reduction of the ore. This "secondary enrichment" was not unusual for Colorado, but the farther the miners sank and drifted, the more complex the ore became, as mentioned earlier. Gilpin County gold miners had quickly confronted this problem of "refractory ores."

The varying character of the ores, especially where zinc was encountered, made treatment difficult and complicated. Smelting had generally failed, as had some other processes including amalgamation and furnaces. High startup costs and reduction charges, plus low saving of the silver content, continued to discourage local mining.

To make matters worse, eastern capitalists had paid "large sums for undeveloped property." Then they sent out "many tons of worthless machinery" and built "large mill-buildings [before] the mines were able to supply them." They were loath to send more. Incompetent managers exacerbated the problems, as they had in other Colorado mining districts. Georgetown's reputation was not a good one.

Raymond retained hope. He examined the most developed mines and found them quite "promising." He also praised some of the mining techniques being utilized. Steam engines hoisted ore up main shafts, Burleigh rock drills eased the workload, and tunnels continued to be driven to strike main lodes at lower elevations—allowing more ease of access and drainage. Of particular interest to him was the "suspension tramway of wire rope" that conveyed ores from the Brown Lode's adit 1,600 feet down the mountain to the mill. Using gravity, the full cars going down pulled the empty ones up the tram. Raymond summarized: "Contrary to the expectations of many engineers, it has worked so far very well."

The Baker Mine also had an "elevated tramway" of 700 feet. Raymond enthusiastically concluded, "[T]his ingenious arrangement is highly to be recommended where the construction of a road-bed might be too expensive."[4]

With little fanfare, in 1869–1870 Georgetown had pioneered two technological breakthroughs that would undergird Colorado mining in the years ahead—steam drills and tramways. The innovations had not yet greatly improved local production, as was typical of many districts that had first used new ideas and faced the usual trial-and-error periods.

The 1870s experienced a technological revolution in mining, which had evolved little over the past 300 years. Machine drills, steam hoists, tramways, dynamite (giant powder, they called it), and new refining methods revolutionized the industry. What proved fascinating about Georgetown in this revolution was the fact that wealthy districts like the Comstock were usually the pioneers. Georgetown was not as rich or as famous as the Comstock, but its mines were very progressive. Georgetown both helped and was helped by these developments.

Two unrelated developments in the 1870s directly helped Georgetown mining—a mining revival on the Comstock and the opening of Colorado's Caribou silver district. These events put silver back in newspaper headlines and in investors' minds. After six years of borrasca, the Comstock's "Big Bonanza" hit full stride, heralding the most productive years ever in American silver mining. The Consolidated Virginia produced $61 million in a nine-year period, and in the banner year—1876—the district yielded slightly more than $38 million. That was a wonderful way to celebrate the nation's centennial and to keep Nevada the number-one mining state.

The Big Bonanza grabbed headlines, created more millionaires than any American mining district before it, and made silver the talk of the day. The way to wealth and fame contained a silver lining. Legends blossomed with this type of wealth as miners, investors, and speculators dived into Comstock stocks and wealthy mine owners paraded their wealth in Virginia City and elsewhere.

Caribou never came close to the Comstock in production or significance, although in its first year an enthusiastic *New York Times* article heralded the discovery: "Wild Excitement among the adventurers of the Territory. The richest silver deposits on the Continent." Every new district wanted to be compared favorably with the Comstock. Caribou—located in Boulder County twenty-five miles and quite a few mountains north of Georgetown—opened in 1870, a new Colorado silver mining district with no connections to the gold bubble. When the Caribou Mine sold to Dutch investors for $3 million in 1873, Colorado silver mining was back in vogue.[5]

Caribou ushered in Colorado's silver decade, and its older neighbor in Clear Creek County became the beneficiary of that development. Georgetown became the soon-to-be state's first "Silver Queen."

One of the innovations Clear Creek miners continued to try in an effort to reduce expenses was tunneling. With claims and mines high on mountainsides, the cost of transporting supplies and equipment up trails remained alarmingly high. Particularly for those with mines just being developed, such expenses added up quickly. As one veteran miner expressed it, "[T]he cost of these roads and trails is no inconsiderable item in the opening and development of a rugged mountainous district like this."

It seemed logical and economical to tunnel in from a lower elevation, strike the lode, and work up and down from there or explore the inner mountain to see what might be found. By cutting costs, such as eliminating the need for "hoisting or special drainage," tunnels would also allow lower-grade ores to be worked. The result was what Rossiter Raymond described as "tunnel fever." Others called it "tunnel excitement," with the number of tunnels started and projected "almost startling."

One of the most famous tunnels was the Burleigh Tunnel on Sherman Mountain between Silver Plume and Brownville. It was the longest one in Colorado in 1876, over 1,900 feet, "in very hard and difficult to blast" rock. Started in 1868, the tunnel was located to explore the mountain: "[I]t was thought it would intersect many valuable lodes that appeared to have no outcrop" rather than strictly undercut any mines. These hoped-for "blind veins" raised expectations in mining districts throughout the West. As Samuel Cushman observed eight years later, however, developments "have not fulfilled the expectations of the proprietors" despite an outlay of $250,000.

The Marshall Tunnel on Leavenworth Mountain, although shorter than the Burleigh, had a similar record. As Cushman summarized, "Like some others, this tunnel has not fully realized the expectations of its proprietors in cutting large and rich ore bodies." Cushman hoped the tunnel would be of some use "as a means of draining and cheaply developing several valuable properties" in the future.

Considering the state of geology and underground surveying in the 1870s and the complex nature of lodes, the tunneling craze held little realistic promise, despite local enthusiasm. The question of who owned what would produce more legal action than ore did. Raymond, in his 1870 report, clearly warned his readers: "A cross-tunnel is likely to be the most expensive of all methods of prospecting. It is run in dead rock, generally hard and costly to excavate, and it cannot furnish ore during its progress to help to bear the cost." He called such tunneling "speculative piracy with which I have no sympathy."

The issue jumped into the pages of the *Engineering and Mining Journal*, which Raymond coedited. When the Burleigh Tunnel encountered a large silver vein, he warned, that success brought about the "imminent danger of a consequent recurrence of the pernicious 'tunnel-fever' which had nearly, we fancied, run its course in the neighborhood and died out."[6]

His comments provoked an irate letter to the *Rocky Mountain News*, berating Raymond for injuring "private enterprise" with "a labored effort to prevent the development of the principal mining region of Colorado." The *Journal* answered point for point. Tunnels, Raymond wrote, had value for transportation and drainage but had to be weighed against the cost of driving them. Not letting up, he condemned "long, expensive and speculative tunnels."

In his 1872 annual report, Raymond further blasted the tunnel fever: "On the whole, the tunnel excitement in Clear Creek County has been productive of no good, but, on the contrary, has been of very considerable detriment to the true interests of the county." Tunneling represented, he summarized, "a premature and ill advised investment of money, which was needed in other and more legitimate mining enterprises." Interestingly, in both his 1872 and 1873 reports Raymond stated that the Marshall and Burleigh tunnels were two of the most promising and energetically developed ones.[7]

Despite local chauvinism, Raymond was right. Tunnels could help mining, but they had taken money, effort, and promotion away from other mining efforts and produced too much expectation. They seemed the simple solution to isolation, distance, and elevation, yet they never panned out for Clear Creek County. As was the case elsewhere, this yearning for a simple, economical answer to some of mining's problems produced more grief than profit.

The machine drills pioneered in Clear Creek mines were much more promising and technologically advanced than the tunnels. The Burleigh Tunnel became especially noted for its use of drills; in fact, Charles Burleigh used them with the intention of introducing the machine drill to the territory and, to a degree, to western mining. Hailed by Raymond as the "most prominent representative of the percussion machine-drills," the power drill promised many benefits over hand drilling, including more speed, lower cost, fewer miners needed, and greater ease of work. Burleigh claimed progress in his tunnel of forty feet per month in 1869, "at least four times as great as could be accomplished by hand labor." The expenditure, he claimed, "is much less than it would have cost by hand labor alone." Each drill only took two men to operate, which Burleigh proclaimed meant savings in wages and other supplies miners needed to single- and double-jack with hand drills and hammers.

Initially, Burleigh had two shifts of eight men working the two drills in operation. Praise was high; the "celebrated Burleigh drill has cut it [the 'very hardest rock'] at the cost of $50 per foot." It was estimated that it would have taken "not less than three years" with hand drilling to accomplish what had been done in a little over a year with power drills.

Troubles with this early model soon appeared, however, because the company was using the large tunnel drill developed for railroad tunnels. It did not adapt well to hard-rock mining and to being subject to an "average 'life' [of five days] in the tunnel without repairs." Too bulky and heavy, the original drills were abandoned, and a more portable style was introduced that was still being used in 1876.

Samuel Cushman felt the failure to achieve any striking success with the Burleigh drill and compressor "in this locality, should not deter anyone from investigating the results of pneumatic drilling under differing circumstances." One could simply look for places where other drills and compressors were successfully operating.[8]

Machine drills, with steam or compressed air, had been around for a generation before Georgetown came into existence. The idea originated in the

United States, but that did not make western miners willing to use them. Conservative by nature, they saw jobs being eliminated and professional skills diminished with the introduction of these machines. They eventually sensed, as well, a health danger from the rock dust. Machine drills, nonetheless, were here to stay.

Meanwhile, Clear Creek mining advanced. Georgetown mining engineer Frederick Eilers wrote the *Engineering and Mining Journal* about that progress: "This silver mining region, which has been struggling for years to emerge from the position of a third or fourth rate camp, had suddenly leaped within the last year to the eminence of the best district in Colorado."[9]

What had caused such a transformation? The discovery of "a number of extraordinarily rich and extensive veins," to start with. Yet the "main spring undoubtedly is a different spirit" that had come over mine owners. Previously, locations were made to sell with sufficient work done only to hold the claims and find "specimens of average value." Somehow, the samples always seemed "high grade." Mines were now being opened and developed—a promising situation. The concerns of the 1860s were finally being addressed.

From strictly a production sense, Eilers was right; 1872 was the first year production topped $1 million. Two years later it surpassed the $2 million plateau. Clear Creek not only challenged Colorado's number-one producer, Gilpin County, but in 1874 even managed—for the first time—to surpass it for one year.[10]

Each mountain around Georgetown had producing mines, although none as rich or famous as those on the booming Comstock. On Republican Mountain, the Dives and Pelican were hailed as bonanza discoveries; they gained fame and then infamy, as will be discussed later. Combined with its neighbors, that mountain "has probably produced more money than any other around Georgetown." Its rival, Democrat Mountain, like its political namesake, wandered "in the wilderness." Several mines were opened in the late 1860s, but many negative comments appeared: "Its [W. B. Astor] production was quite large for a time, but work was afterwards suspended." High on this mountain the miners lived in hope: "No very great activity upon this class of lodes can be expected until the completion of the railroad, giving cheaper transportation to the coal regions."[11]

The Terrible Mine was the feature on Brown Mountain, "one of the leading producers in the county ever since silver mining became prominent." Sold to English investors in 1870, it produced steadily for five years until it ran into an apex dispute and work was suspended. When work resumed, the Terrible became part of the Colorado United Mining Company. The typical pattern unfolded here. Discovered by prospectors and sold to outside investors who brought the mine into production, the mine became involved in a mining dispute and was finally combined with other mines. By the late 1870s, although over 100 miners worked for the company, leasers also mined in sections of the property.

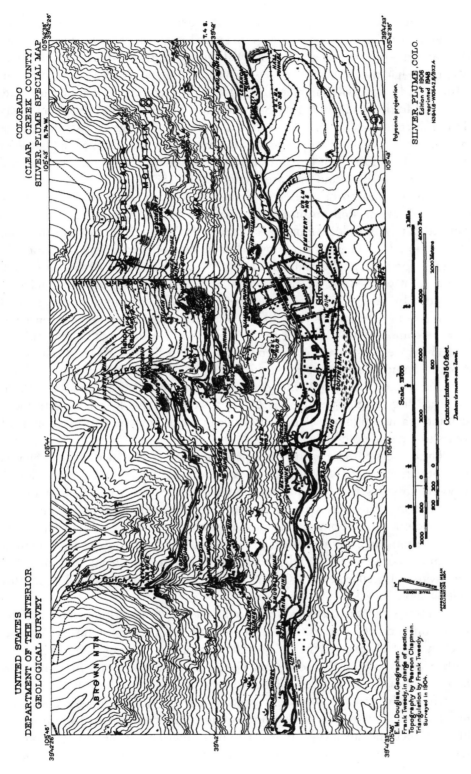

Many of the best mines "around Georgetown" were actually in Silver Plume. Special map of Silver Plume, USGS topographic map, 1906. Christine Bradley Collection.

Griffith Mountain promised much, with "extremely rich ore" at the surface to tempt the prospector. It never matched that early promise. Sherman Mountain—named after General William Tecumseh Sherman, hated in Georgia—had the Burleigh Tunnel with its intriguing tale. McClellan Mountain, also named for a Civil War general and 1864 Democratic presidential candidate George McClellan, included the Stevens Mine. Noted for the longest tramway in the district—over 1,000 feet—the mine claimed George Armstrong Custer as one of its owners. Poor Custer had no more luck here than at the battle of the Little Big Horn. Just beyond sat Kelso Mountain, hampered by poor and limited wagon roads and low production in the 1870s.[12]

Some of these properties were highly developed, with 1,000-foot adits, extensive levels, shafts hundreds of feet deep, and machinery to match. Unfortunately, most local mines lacked these attributes. Of Democrat Mountain it was noted, "The entire mountain must have yielded nearly a million in silver"—for a decade's production. The W. B. Astor Mine offered a perfect example of what went wrong. Owned since 1868 by a group of Cincinnati investors, by July 1877 only six men were working it. The local paper assessed the situation: "The mine is yet in its infancy, not having been worked extensively, but the company proposes to employ a large force of miners this season."[13]

There were too many comments such as "but little work has been done" and the mine "has not been worked steadily." When total production was listed at $40,000 or $50,000 a year, a mine was considered a small producer. Another discouraging trend in the 1870s was the fact that leasing became more common, a sure sign of lack of confidence in mines anywhere from a couple of years to a decade old.

The owners of these properties represented a varied lot, with different levels of motivation. Some were well-to-do and famous, whereas many were locals who might be described as small mine owners. Some came from the East, others from Europe, but most were from the West. A few had mining experience, more were mining rookies, and most invested because a gold or silver mine seemed almost like having the ore in the bank. Nearly all of them had caught a case of gold or silver fever and expected to make a fortune from their mines. In the words of Shakespeare, "Gold? Yellow, glittering precious gold . . . this yellow slave."

All this activity had changed the local scene. *Colorado Miner* editor E.H.N. Patterson had been one of those "irrepressible prospectors" who had arrived back in 1860. He recalled the valley at that time as heavily timbered and covered with beaver dams to the point that he had to scramble along the mountain base over boulders and fallen timber to get to the head of the valley.

The timber and beaver were gone, the creek was no longer clear, the mountainsides were scarred with mine dumps and crisscrossed by trails, and litter marked the advance of industry and population. At the moment, only a few seemed concerned. After all, the country had been opened, natural resources had been exploited, settlement had taken root, and profit had been made—all to the benefit of America and Americans.[14]

In 1876, amid statehood celebrations, veteran Georgetown miner and soon-to-be newspaperman Samuel Cushman published a lengthy pamphlet, *The Mines of Clear Creek County.* He summarized what had happened during the past few years.

> The facilities for the production and treatment of ore and the supply of skilled labor have greatly improved within a few years, furnishing the requisite conditions of success that were formerly lacking. The result of these changes is witnessed in the steadily increasing prosperity of the mining business, foreshadowing what that business will yet become when supplies and labor are still cheaper, when labor-saving appliances are more generally adopted, and when a still larger measure of success is attained in the treatment of all classes of ores.[15]

Cushman understood what had transpired and clearly saw what was yet needed.

Clear Creek County silver production had definitely improved. The $1 million level held true for the rest of the decade, although the $2 million total was rare. That range was maintained, in fact, for the rest of the century. Unfortunately, not everything was turning out so well for Colorado's Silver Queen.

For one thing, the *Georgetown Courier* complained: "It must be admitted by every candid judge that a considerable portion of our mines are not systemically worked." Others were badly managed. Without naming names, the editor wanted "systematic mining." That meant a well-considered plan of working, expert supervision, and long leases, but it did not infer "a large work force or administration." Those same thoughts echoed from numerous western mining districts.[16]

Litigation over claims and veins highlighted the history of many mining districts; Nevada's legal battles, for example, were famous. Georgetown proved no exception. The 1870s witnessed a fight between two highly productive mines on Republican Mountain about two and a half miles west of Georgetown—the Pelican and the Dives. The dispute centered on who held the apex of the vein.

The apex "law," or "extralateral right," was part of the 1872 National Mining Law. This right allowed the owners of a claim the exclusive right to mine a vein if the apex, or highest point, occurred within their property. The owner could follow the vein's downward course beyond the property's sidelines but not beyond the end lines unless he or she purchased the neighboring claims. If claims had emulated the orderly seams described in textbooks, this plan might have been practical. In reality, such veins seldom existed in the mining world. Veins and ore pockets went in every direction and often surfaced in other claims. According to James Hague, "Diverging, converging, broken, crooked or otherwise nonconformable veins" bedeviled everyone. With geology in its infancy, how could offshoots of a main vein be distinguished underground or on the surface? As the old mining saying went, "[Y]ou could not see beyond the pick at the end of a mine."

Lawsuits resulted. Although it was intended to simplify, "establish, and confirm a clear and unassailable title," the apex law virtually did the opposite. Predictably, such litigation nearly always occurred between the owners of a rich mine and those who owned less valuable adjoining claims, although two mines could battle for complete control of a deposit. Looking back over thirty years of apex woes and riddles, distinguished mining engineer James Hague told an audience at the 1904 St. Louis Exposition that the apex law "brings forth nothing but confusion and contention, with everlasting and costly litigations."[17]

The Pelican-Dives conflict classically exemplified exactly what Raymond had pointed out earlier. It was Colorado's first major fight over the apex law, and before it ended Georgetown's reputation suffered, mine owners and their mines were nearly ruined, violence replaced legal maneuvering, and some of the big names in Colorado mining were dragged into the mess.[18]

The Pelican outcropping had been discovered in 1868, but actual mining did not start until December 1870. The owners (Eli Streeter and Thomas and John McCunniff) developed their mine carefully and slowly over the following two years, uncovering some rich silver deposits. Georgetown's *Colorado Miner* called it "the richest mine in Colorado," with ore running between 800 and 1,000 ounces of silver per ton. The Pelican's reputation grew over the next year. On November 28, 1874, the weekly edition of the *Miner* reported that 100 men were busy at the mine, but more exciting was the fact that immense amounts of ore had been exposed.

The Dives claim had been staked in May 1869 by Thomas Burr, who the next July conveyed a half interest in the property to upcoming Georgetown entrepreneur William Hamill. Short of money, in September 1870 Hamill gave one of the town's young lawyers, John McMurdy, a deed to his share for $10,000. McMurdy purchased Burr's interest the next May and set about to develop the Dives Mine.

Within a few weeks McMurdy went back East and organized the Perdue Gold and Silver Mining and Ore Reducing Company, conveying to it part of the Dives property. The Dives failed to receive the attention the Pelican did over the next few years, but it was a good property. The *Colorado Miner* stated in late 1873 that the "mine is opening up magnificently," with a two-foot-thick pay vein. Meanwhile, William Hamill came back as manager, and some prominent investors came aboard—including Civil War hero Senator John Logan.

Unfortunately for all concerned, parts of their claims overlapped, opening the door for everything that followed. On December 27, 1873, Streeter and McCunniff filed a complaint of trespass against the Perdue Company. Faulty surveys, conflicting testimony, and disputed discovery dates all made for heated arguments. Neither party allowed the other to inspect the drifts in which they were working, which heightened charges that the parties had removed ore from their opponents' property. In April the Perdue Company filed a counterstatement. Throughout the rest of the year legal proceedings wrangled on in courtrooms and lawyers' offices.

Neither side terminated mining in areas close to the disputed section, leading to new lawsuits. Tension mounted as the days passed. Even at the start of the feud the *Georgetown Courier* reported that the courtroom was a "walking arsenal."

After a change of venue, one of Colorado's respected lawyers and judge of the Third Judicial District in El Paso County, Moses Hallett, heard the arguments. In April 1875 he decided two pending cases in ejectment in favor of the Pelican, but he also lifted the injunction against the Dives, allowing the miners to work.

Hamill prepared to resume operations, but on the night of April 24, armed men from the Pelican invaded the Dives and seized control of underground property they had never worked previously. Asserting that the land was Pelican property and previously unoccupied, they claimed they needed to be armed to retain possession. Matters quickly worsened.

Both sides hired armed guards, who were placed on the surface and inside the workings. Each set appeared content to keep the other at bay, an attitude that prevented violence. Two deputy sheriffs were also stationed nearby, but they could have done little if shooting started. Visitors were denied access to the workings, but rumors crept out that Pelican miners had found a rich vein in the disputed property.

To many Georgetown folks it seemed that old-timers (Pelican) were pitted against newcomers (Dives). This situation, of absentee owners with money versus hardworking miners who had helped open and develop this land, created further ill feelings.

Failing to get Hallett to grant them legal recourse to remove the Pelican people, the Dives turned to Judge Amherst Stone in Central City. Stone promptly issued an injunction on May 18, and acting sheriff Presley Baily delivered a copy to the Pelican defendants. Baily got nowhere; he was informed that the Pelicans would defend their property, and before leaving he was shot at. Unable to seize the property without bloodshed, the sheriff attempted to starve the men out by posting guards at all the mine's openings. That strategy failed when the Pelicans somehow managed to receive supplies.

The Pelican owners had hired Henry Teller, one of the territory's leading mining lawyers, and Baily and Hamill traveled to Central City to see him. He refused to transfer the disputed property to them. Frustrated here, Baily also made no headway at the mine. Finally, on May 20, he removed his men.

At this point a controversial, hotheaded former member of Quantrill's band during the Civil War, Jackson Bishop, decided to resolve the interminable legal wrangling. His involvement came about because he had a lease on some of the disputed territory. On the afternoon of May 20, 1875, the enraged, "frenzied" Bishop encountered banker Jacob Snider, one of the Pelican owners, with whom he had been on bad terms because of a dispute over a mining lode. He rode Snider down, then shot and killed him in front of "several hundred witnesses" who had raced outside after hearing the commotion. No one did anything, and Bishop escaped. Fear reigned. Unfortunately, Bishop was never arrested and brought to trial.

The violence that people had been dreading for weeks had arrived. Heated words, threats, impatience, disgust, and hatred had produced a not unexpected result. For awhile lawlessness seemed to triumph over abiding by the law, particularly concerning Snider's murder. To some locals the worst thing about the incident was the bad reputation the town acquired as a result. Opinion about what had happened, who had or had not done what, plots and conspiracies, and guilty parties' involvement lingered on for a generation.

The tension and panic of the moment can be seen in telegrams sent to Teller, who desperately tried to understand the situation. These four arrived in rapid order on May 24.

> Baily summoning another posse to take possession.

> Your dispatch at hand. When every hope deserts me will surrender not before.

> To save more bloodshed see what arrangement you can make with opposing counsel.

> The mine will be taken & more than one poor fellow will be killed & more of your clients may fall. For God's sake prevent further bloodshed.

Poor Teller, only a few miles away, had no idea what was actually transpiring.[19]

Both sides now realized that the legal wrangling, armed camps, violence, and bad publicity had hurt the district, the community, and mining in general and might divert capital away from the region. Positions were modified, reason took hold, and the situation started to return to normal.

It was several years before all the uneasiness and insecurity, plus the original land dispute and lawsuits, were finally resolved. The Pelican owners—underfunded—admitted in the spring of 1876 that they had spent half a million dollars defending themselves. With this outlay undermining them, they agreed to discontinue legal proceedings and divide the disputed land. Stockholders, Snider's heirs, and others filed lawsuits, and the cases dragged on into the 1880s. Ultimately, under Hamill's direction, the two properties were merged, and the Pelican and Dives Mining Company emerged.

For the company, sadly, the days of rich production were past. Lawyers on both sides came out better than the miners and owners. Georgetown's reputation suffered, and so did its mines. Unfortunately, the aftermath occurred at a time when the Leadville excitement started generating interest in an even bigger

Colorado silver bonanza. The complexities of the late-nineteenth-century mining industry came to the forefront during the Pelican versus Dives dispute. Both sides argued that they possessed the apex, discovery right, territorial proprietorship, and other aspects of their mines; and a number of young lawyers learned about legal frustrations through the resulting lawsuits. The struggle also displayed the complications that developed when eastern finance entered the western economy. Westerners needed the capital, but they did not necessarily like everything that came with it. The use of toughs by both sides also typified some of these heated mining disputes, in which violence usually resulted.

Not unexpectedly, much of the profit from these two mines' bonanza years had been swallowed by the litigation. In the end, no one really gained, and Georgetown never fully recovered its once-held prominence as Colorado's Silver Queen. Further, this was not Georgetown's only apex case. William Hamill became embroiled in a variation of the Pelican-Dives conflict when he took on the Terrible Mining Company on Brown Mountain. Initially, as a small operator, he and others had purchased property next to productive mines. A predictable next step followed: individuals and corporations became entangled in time-consuming, lengthy legal proceedings.

The Terrible, a British-owned company, was a profitable operation when Hamill's miners opened a drift from the Silver Ore Mine into the Terrible workings in 1874. An apex lawsuit followed in 1875. Joined by Colorado smelter man and master politician Jerome Chaffee, who also owned claims on the mountain, Hamill and his colleagues forced the British to concede and, in 1877, to purchase the neighboring claims. That avoided time-consuming, costly litigation. Chaffee and Hamill came out of the transaction with shares in the new consolidated company. Although the case was slightly different from Pelican-Dives, the result was the same.[20]

Caribou faced similar troubles with mining properties, but they proved less costly, violent, and time-consuming. Conversely, Nevada's Comstock became infamous for its mining disputes, fights, and costly litigation. Georgetown fell between these two extremes, but in all such cases the troubles made investors nervous and did not help the mines' or the districts' reputations.

An older concern had also not been completely resolved—the issue of profitable ore reduction. Hill's Boston and Colorado Smelter pointed the direction, but even it had problems with low-grade ore—which predominated in every western mining district. Noted mining man George Hearst observed: "Here's to low grade ore and plenty of it." Of course, his famous Homestake Mine in the Black Hills of South Dakota contained such ore, and it produced for 125 years. Most mines were lucky to have profitable production for a decade. Unfortunately as well, many owners and stockholders loved high-grade ore and did not seek long-range operation.

Low-grade ore required processes to mill and smelt that were as complicated and costly as those needed for high-grade ore, with, unfortunately, fewer ounces of gold and silver rewarding mine and mill owners. Many districts died early deaths because they lacked a reduction works to handle low-grade ore, which could not absorb long-distance transportation costs.

In the 1860s Georgetown miners had grappled with this problem and the equally significant issue of what process worked best, without noticeable results. They held high hopes for the Stewart Silver Reducing Company, which started to build a smelter at the lower end of town in 1870. J. Oscar Stewart built the mill and was given credit for the "most persistent effort to work the ores of Clear Creek county in Georgetown." A "wondrous mechanic," Stewart fabricated several previously patented inventions, with the result that he became embroiled in several patent suits.

The planned process included using stamps to crush the ore, then roasting and cholodizing to convert the sulphide silver into a chloride that could then be amalgamated. Someone counted 145,000 bricks and 175,000 feet of lumber that would go into the various buildings. Everything worked successfully until January 1872, when the works burned.

The works were rebuilt and opened in September that same year, and the company continued to tinker with the process. Stewart hired David Brunton, a young mining engineer recently graduated from the University of Michigan, in the summer of 1875 to make the process work. He did make the works profitable, and the situation seemed more promising than ever. Modestly, years later Brunton recalled his efforts: "It was passing out of the experimental stage and I was fortunate enough to come in just as it was possible to bring the process into profitable operation."

With amazing bad luck, the plant burned again in December 1875, and Cushman noted it was rebuilt the next year. Local pride felt the works had been a success, and it probably was the most successful district smelter in Colorado, producing about $1.5 million in bullion.

Still, the company had experimented with various processes before finally settling on one that worked. Suffering two disastrous fires also did not help. Although it will never be known how profitable the operation might have been, considering all its troubles it could not have returned a handsome dividend. Brunton observed, though, that the company "immediately began making money, a change that soon attracted the attention of the other mill-owners in the district."[21]

Brunton continued his reminiscences. "As soon as it was known that the plant was a success," H. Augustus Taylor, owner of the Clear Creek Reduction Company, "offered me a much larger salary than I was getting." Brunton designed and directed construction of an amalgamation and leaching mill that was a success. Eventually, by developing a leaching process, the young engineer made amalgamation unnecessary. His mill became the first that could profitably work low-grade ores. A visitor in 1877 gave the plant high praise, saying it "is to Georgetown what Professor Hill's works are to Gilpin county, as it is the largest and by far the finest mill for the reduction of silver ores that I know of in Colorado." Further, and more crucial to the owners and miners alike, "it is managed on business principles and making money for its owners."

Perhaps less easily discernible at the time, Brunton represented the arrival of a new breed in Georgetown mining—the U.S.-trained mining engineer. They came in larger numbers now, as the new American schools of mining started to turn out graduates. Within a few years, however, many would cross the mountains to Leadville, taking with them the experience they had gained from having worked in the first Silver Queen. Brunton left after "about a year and a half," going to nearby Caribou in Boulder County and its silver mines and continuing a career that would take him throughout much of the mining West.[22]

The district also had concentrating and sampling works. Hill's Boston and Colorado Company built a sampling works in 1874, then sold it to local residents; it then became the Church Bros. Sampling Works. Such operations provided two services for the miners: a place to take ore to be tested and to then either sell it to the works or know what the ore was worth prior to transporting it to a smelter for reduction. Hill's smelter bought "the larger part" of the sampling works ore.

A concentrating works simply removed "by mechanical means the lighter and less valuable portions of ore." For example, six tons of ore might be reduced to one or two tons, which allowed for less costly shipping. Less expensive to build than sampling works, concentrating works partially solved some of the reduction problems Clear Creek mines faced.[23]

Georgetown, which had become the center of experimentation on silver, paved the way for Colorado's future silver districts. It hit upon a smelting method that seemed successful. Based on crushing, chloridizing, roasting, and amalgamating, the process evolved into one that adapted to most of the district's ores. Obviously, one process would not serve them all, just as Nevada's Comstock methods had not totally matched Colorado's ores. Forebodingly, too much money and time and too many hopes had been dashed along the way.

The average small mine owner who could not afford to build his own reduction works had to rely on someone else's efforts. The district, however, never gave up its dream of having its own major smelting works, similar to what neighboring Gilpin County had with Hill's works. That, to them, would be the elixir of life, the answer to making Clear Creek County an ongoing bonanza.

Another benefit for local mining came with the appearance of skilled hard-rock miners. Unlike placer operations, which could be mastered in a relatively short time, when mining burrowed underground it entered a complex, technical world. It took experience to drill, timber, hoist, blast, drain, and develop a "nose for ore." Fortunately for all concerned, Cornish miners arrived in increasing numbers in Gilpin County and spread out from there into Boulder and Clear Creek counties.

Coming from Cornwall in southeastern England, the miners had a heritage of mining that stretched back to Roman times. Mining ran in their blood, and they indeed possessed a nose for ore. With their famous tin and arsenic mines declining, the Cousin Jacks, as they were nicknamed, emigrated throughout the world. Many came to North America, moving from Michigan's copper mines to Missouri lead mines and on to Colorado and the West. They brought with them their centuries of honed skills and their equipment, such as the Cornish pump. Not only did the Cornish contribute, but, having mined for over a decade, a nucleus of local miners was also experienced in underground mining.

As the decade neared its end, Georgetown reached its apex as a silver mining district. Frank Fossett, Colorado's most prolific and one of its best 1870s mining reporters, estimated that "probably two thousand men are directly engaged in mining, milling, and hauling ore in Clear Creek county, or in prospecting." He went on to quote from an undated *Colorado Miner* that, despite its Victorian sentimentality, paid an honest tribute to the miners who had opened and developed Georgetown's mines.

> For a period of about sixteen years, that enterprising individual, the "honest miner," has prospected and dug for the precious metals in our county with that energy and tenacity which is a distinguishing characteristic of "miner men" and, to some extent, is born of the circumstances in which he is placed. He has lived hard and worked harder, and with an undaunted brow has often faced the bitterest and sternest realities of life. . . . [H]e has toiled incessantly for years without taking out a "red," but his faith still continues unshaken and his perseverance unimpaired. The brave and persistent miner has done all this and much more; not with the magic wand of an eastern fairy, but with a striking hammer, weighing from six to eight pounds, and other implements necessary to his vocation.[24]

For this the miner received $2.50, $3.00, or perhaps $3.50 a day, if he toiled in a wet mine, traveled to work on his own time no matter how far his work lay from the mine's portal, and lived a dark, dangerous, labor-intensive life once he reached the breast of the drift. By the flickering light of his candle— a commodity he dared not waste because he only had enough for one shift (walking out in the dark was no fun)—the hard-rock stiff drilled, blasted, mucked, and trammed the ore to the surface.

They lived a transitory life. Miners tramped down the mountain for as many reasons as there were individuals. In their opinion theirs was a skilled profession, and they took pride in it. Even high grading reached an artform when miners supplemented their income by stealing ore. The Cousin Jacks seemed especially adept at this. As one frustrated mine manager said, "You can't work a mine without them, you can't make a profit with them."

As 1877 raced by, local expectations soared, perhaps higher than ever before. The Colorado Central Railroad was nearing the district. The iron horse reached Idaho Springs in June, and on August 1—the first anniversary of statehood—it chugged into Georgetown. Appropriate ceremonies, gold and silver spikes, laudatory speeches, and 8,000 happy people greeted the "savior." The railroad promised manifold blessings for townspeople and miners. Those dreamed-of benefits would now become reality, or so the optimistic anticipated. A lower cost of living and lower shipping charges, ease and comfort for investors coming to see the mines, tourists, the end of transportation problems, diminished isolation, and a brighter future at hand exemplified only a few of the promised or prayed-for blessings of this wonder of the age.

The *Colorado Miner* greeted the new year with a special eight-page edition on January 12, 1878, with drawings of leading mines and mills on page 1. Optimism and hope permeated every page. The future looked bright indeed. However, a long letter in the issue discussed the silver question. The metal's price had steadily slipped downward, from $1.33 to $1.15 an ounce. Equally as bad from the miners' viewpoint, the government had "sneakily" stopped coining silver back in 1873. Silver miners everywhere were becoming aware of this "crime" and were well aware that the ore price had declined alarmingly.[25]

What Georgetown did not know amid the fun of the New Year's celebration and the exhilaration of the railroad's recent arrival was that its days as Colorado's Silver Queen were numbered in months, not years. Uneasy rested the crown, and a challenger had already seized the attention of Coloradans and other investors.

NOTES

1. Mining figures in *Rocky Mountain News*, November 8, 1869, 1.

2. Charles Henderson, *Mining in Colorado* (Washington, D.C.: Government Printing Office, 1926), 109, 122; Grant Smith, *The History of the Comstock Lode, 1850–1920* (Reno: Vanier Graphics, 1980 reprint), 292; Clark C. Spence, "The British and the Colorado Mining Bureau," *Colorado Magazine* (April 1956): 81–85, 91–92; Old quoted in Clark C. Spence, "Robert Orchard Old and the British and Colorado Mining Bureau," unpublished M.A. thesis, University of Colorado, Boulder, 1951, 2, 44–46, 109–116.

3. James W. Taylor, *Report on the Mineral Resources of the United States East of the Rocky Mountains* (Washington, D.C.: Government Printing Office, 1868), 8–11.

4. Rossiter W. Raymond, *Statistics of Mines and Mining* (Washington, D.C.: Government Printing Office, 1870), 368–378.

5. For the Comstock's Big Bonanza, see Smith, *Comstock Lode*, chapters 16–18; Rodman Paul, *Mining Frontiers of the Far West* (New York: Holt, Rinehart and Winston, 1962), chapters 4 and 5. For Caribou, see Duane A. Smith, *Silver Saga* (Boulder: Pruett, 1974), chapters 1 and 2. Also, *New York Times*, September 12, 1870, 1.

6. *Engineering and Mining Journal*, April 18, 1872, 297; quotes from May 9, 1872, 249; December 9, 1873, 381; *Rocky Mountain News*, November 8, 1869, 1 (miner quote), 4; Raymond, 322–323; Samuel Cushman, *The Mines of Clear Creek County* (Denver: Times-Steam, 1876), 4, 33–35, 57–59.

7. *Rocky Mountain News*, April 27, 1871, 4; Rossiter W. Raymond, *Statistics of Mines and Mining* (Washington, D.C.: Government Printing Office, 1872), 322–26;

Rossiter W. Raymond, *Statistics of Mines and Mining* (Washington, D.C.: Government Printing Office, 1873), 260–270; Cushman, *Mines,* 4, 33–35, 57–59.

8. Rossiter W. Raymond, *Statistics of Mines and Mining* (Washington, D.C.: Government Printing Office, 1869), chapter 67, discusses machine drills and nearly a dozen others; Burleigh-related quotes are also from this source. His 1871 report has an entire chapter (chapter 20) on the Burleigh drill in which he concluded "it has never been successfully pushed in the West"; Rossiter W. Raymond, *Statistics of Mines and Mining* (Washington, D.C.: Government Printing Office, 1871), 487–492. Also Cushman, *Mines,* 34–35; *Rocky Mountain News,* November 8, 1869, 4.

9. *Engineering and Mining Journal,* September 3, 1872, 155.

10. Henderson, *Mining in Colorado,* 109, 122.

11. *Colorado Miner,* 1868–1869.

12. Frank Fossett, *Colorado* (New York: C. G. Crawford, 1880), chapters 14–16, especially chapter 14; Cushman, *Mines,* 30–56, 64–66.

13. *Georgetown Courier,* June 14, 1877, 1.

14. Patterson talked about the "old days" in the *Colorado Miner* [Georgetown], January 12, 1878, 8.

15. Cushman, *Mines,* 4.

16. *Georgetown Courier,* June 14, 1877, 1.

17. For the apex law, see James D. Hague, "Mining Engineering and Mining Law," *Engineering and Mining Journal* (October 20, 1904).

18. The discussion of the Pelican-Dives conflict is based on Liston E. Leyendecker, "The Pelican-Dives Feud," in *Essays and Monographs in Colorado History* (Denver: Colorado Historical Society, 1985); Duane A. Smith, *Henry M. Teller* (Boulder: University Press of Colorado, 2002). Also *Colorado Miner* [Georgetown], January 14, 1873, 2; October 10, 1873, 2; November 28, 1874, 1.

19. *Colorado Miner* [Georgetown], January 14, 1873, 1; October 10, 1873, 1; E. Naylor to Teller, Wm. Rockwell to Judge Forsline, May 24, 1875, Henry Teller Papers, University Archives, University of Colorado, Boulder.

20. Christine Bradley, *William A. Hamill* (Fort Collins: Colorado State University Press, 1977), 7–10.

21. *Engineering and Mining Journal,* October 5, 1869, 223; September 6, 1870, 148; June 18, 1872, 396; building figures in Cushman, *Mines,* 94–95; *Colorado Miner* [Georgetown], September 23, 1869; Stewart quoted in Raymond, *Statistics of Mines and Mining,* 1871, 320–321; David Brunton interview in T. A. Rickard, *Interviews With Mining Engineers* (San Francisco: Mining and Scientific Press, 1922), 70.

22. Visitor quoted in *Rocky Mountain News,* April 17, 1877, 3; Brunton interview in Rickard, *Interviews,* 70–71, 92–93.

23. Cushman, *Mines,* 94–95, 99, 100–101, 103; quotation in Rossiter W. Raymond, *A Glossary of Mining and Metallurgical Terms* (Easton, Pa.: American Institute of Mining Engineers, 1881), 24; *Rocky Mountain News,* April 4, 1877, 3; Robert Spude, correspondence with authors, July 23, 2002.

24. Fossett, *Colorado,* 365–366.

25. *Colorado Miner* [Georgetown], January 12, 1878, 3.

British Mining

"Rule Britannia, Britannia rules the waves" was England's rallying cry. Britannia ruled more than the waves; its investors were crucial to western mining development.

Georgetown offered no exception. The community's hopes, however, far outpaced success in luring British money to the valley.

Robert O. Old home at the base of Democrat Mountain, Georgetown. Courtesy, Denver Public Library, Western History Department, X-1223.

Robert Old tried for years to attract British investors, and so did others. When the money did arrive, it did not always have the positive effect for which investors had hoped. Two classic cases hurt Georgetown's reputation among the English. The Colorado Terrible Lode Mining Company, organized in 1870, found itself in litigation in 1875 when William Hamill's Silver Ore miners broke into the Terrible Lode mine. The case concluded with the British purchasing that property and several others. More than one observer accused Hamill of deliberately pushing development to create a dispute so he could sell his property. The company, nevertheless, paid dividends. Its red streamer, emblazoned with the silver letters "Terrible," floated over its property for all to see.

The Pelican-Dives turmoil over mineral deposits caught other British investors. Their hopes of dividends went down the mine. Here and elsewhere, the English frequently complained of misrepresentation by promoters and of "salted" mines.

Clearly, frauds did occur, and even in legitimate transactions expectations were not always matched by dividends. Westerners needed those investments, however, so they continued to offer enticements, and English investors seemed more than willing to gamble on winning high stakes.

6 Georgetown, 1878
One Year in the Life
of Colorado's Silver Queen

Financial conditions in Colorado looked promising on January 1, 1878, because most commercial enterprises, except manufacturing, had prospered during 1877. The newly admitted state's laboring population was increasing, its railroads and wagon roads were growing, and its meats, dairy supplies, and agricultural products generally showed development. Stockmen were prospering, for horse, cattle, and sheep herds continued to expand as they had for the preceding five years. Markets improved steadily, with no disastrous losses, whereas stock grades improved every year. Colorado's beef was as highly desirable as any in eastern markets, and increasing numbers of buyers found its wool acceptable. All of this was the result of production and cheap competitive transportation to eastern markets, which gave Colorado a leg up over rival states.

Crops expanded as growers learned how to combat grasshoppers, the only serious pest in the area. Farms were in demand, and railway companies sold more land; preemptors were taking up claims, and it was not difficult to sell easily watered lands. New irrigation ditches appeared, and those already dug were enlarged and extended. The *Rocky Mountain News* predicted that Colorado's agricultural product would increase by 50 percent in 1878.

Railway construction had boomed during 1877, with the Colorado Central showing the greatest growth. Its seventy-two-mile extension from Longmont to Cheyenne had opened the Cache la Poudre and Thompson valleys to faster commerce. The line's extension from Floyd Hill to Georgetown was laid, and the five-mile addition from Black Hawk to Central City was graded and nearly ready for track. Actually, all the railroads in the state had been busy, so that Colorado now boasted 1,200 miles of track.

Unfortunately, Colorado's manufacturing had not kept pace with other parts of the state's industry, for although its foundries and ironworks produced engines, the stock had to be imported from the East. The state possessed plenty of flour mills, but it had little else.

Nevertheless, a growing number of miners and mining settlements had led to an increase in that industry. The miners' attitudes seemed to have changed, for they appeared less inclined to move on to greener pastures, and they operated their producing properties in a more economical and systematic fashion—determined, as it were, to settle down and make a go of mining. Colorado's output of gold and silver was increasing. Its gold production—$3,151,277 in 1877—was a gain of $321,400, while the silver output of $3,197,861 equaled $401,200 more than the year before. This heavy growth in 1877 was due largely to Clear Creek and Gilpin counties, the state's two leading producers.[1] Silver was Clear Creek County's prime product, and indications looked good for passage of a silver bill designed to put the "Dollar of Our Daddys" back into circulation.

The federal government had removed the standard silver dollar from its coinage lists by means of the Fourth Coinage Act, generally called the "Demonetizing Act," of February 12, 1873.[2] This legislation had been supported by eastern bankers and merchants who wanted to place the country on the gold standard. The money men believed the measure would ease America's dealings with Western European countries such as England, which were already on the gold standard. Another reason easterners gave was their belief that the country did not need the inflated greenbacks issued during the Civil War. Such currency, they believed, should be retired in favor of specie or hard money, preferably gold. These actions, together with overexpansion of railroads, led to a financial panic in September 1873 that led to a four-year depression.

Meanwhile, in 1876 Representative Richard Bland of Missouri enlightened many of his congressional colleagues about the "Crime of 1873," as he called the Fourth Coinage Act, and he introduced the Bland Bill to restore unlimited coinage of silver at the traditional ratio of sixteen parts silver to one part gold. Senator William B. Allison of Iowa, chair of the Senate Finance Committee, ushered the bill through the Senate, seeing that much of its inflationary intent was removed.[3] As 1877 glided to a close, the Bland-Allison Bill was nearly ready for passage.

Residents of Colorado's chief silver camp, Georgetown, kept an eye on the bill's progress through the Senate, fearful that President Rutherford B. Hayes would veto it. Obviously, the bill's passage would lead to a greater demand—and a higher price—for the white metal.

While eastern newspapers urged the bill's defeat, those in the West—including Georgetown's *Colorado Miner*—championed its enactment; the latter papers took Secretary of the Treasury John Sherman to task for his contractionist views.[4] Anticipation grew as the *Colorado Miner* covered the bill's progress throughout January and February. Readers took heart when the newspaper reported that passage of the bill would provide confidence in Colorado's resources and give stability to its production, thus creating an inducement for easterners seeking profitable investment to venture more money in the state than ever before.[5]

On February 16, 1878, when the "Silver Bill" passed the Senate, the people of Georgetown celebrated and fired iron and brass cannon near the Pelican Mill in honor of the congressmen who had supported the "Dollar of Our Daddys."[6] General F. J. Marshall, a prominent mining operator; B. F. Napheys, a mill man and president of Georgetown's Board of Trade; and E.H.N. (Sniktau) Patterson, editor of the *Miner*, responded to the Senate's action by wiring Colorado representative Thomas M. Patterson: "Hurrah for the Senate. We are firing four hundred and twelve and a half guns in honor of the supporters of our daddy's dollar. Great rejoicing and jubilation."[7] Patterson had opposed the bill in its final form because he did not concur with the Senate amendments.

The bill became the Bland-Allison Act, over President Hayes's veto, on February 28, 1878.[8] It ordered purchase of $2 million to $4 million worth of silver a month, at the market price, for coinage into dollars.[9]

The people of Georgetown rejoiced, even though the free coinage clause in the original bill had been stricken from the final act. Nevertheless, there was hope that in time it would be revived and enacted into law.[10] Meanwhile, the director of the mint had arranged to have the mints run at full capacity to produce the new silver dollar. As working dies were being readied, the bullion was processed for coinage into dollars. The government anticipated little delay and estimated that by mid-April silver dollars would be produced at the rate of $3.5 million per month.[11] Such prognostications were correct, for in the middle of April Billy Tobin advertised that he had the new silver dollar and would "show it to customers of the O.K.," a store selling "Gent's Furnishing Goods."[12]

Once the Bland-Allison Act was a reality, Georgetown's inhabitants turned to matters closer to home by petitioning the county commissioners to remove the powder magazines from the mountainsides west of Alpine Street to an area where, if they exploded, no harm would be done to the town's buildings.[13] Community activists considered the mining camp's physical as well as financial survival.

As in most American municipalities, local politics played a part in Georgetown's life. On March 16, 1878, the *Colorado Miner* announced the results of voter registration in the town's three wards. The first ward reported

247 registrants, the second had 218, and the third reported 356, for a total of 821 voters—indicating that voters in Georgetown made up about 16.5 percent of the town's population. Persons who had not registered but who were entitled to vote had to appear personally before the board on the Tuesday preceding the election to have their names entered in the voters' lists.

The Democrats held the first convention and unanimously nominated Jacob Fillius, the incumbent police judge, as a candidate for a second term. A committee was dispatched to notify him of his nomination. Convention members cheered Fillius when he arrived shortly afterward to make a short speech, thanking the group for endorsing his official actions over the preceding two years. If elected, he pledged to continue to devote his energy to "the general welfare of the town and its people."

Democrats then separated to nominate selectmen from each of the town's three wards: first ward, C. E. Wyman; second ward, P. McCann; and third ward, J. W. Forest. All candidates promised to work for Georgetown's best interests.[14]

The following Saturday evening, March 16, 1878, Georgetown's 140-member Greenback Club gathered at Old's Hall to elect officers and to nominate a municipal ticket. As their chairman they elected Robert O. Old, who delivered a Greenbacker's speech on the country's finances. The membership voted for a candidate for police judge, choosing Jacob Fillius by twenty-one votes over eight for J. Van De Voor. Fillius's nomination was made unanimous.

Members cheered when Fillius appeared before them. He told the group he was flattered by their decision but reminded his audience that he had accepted the Democratic nomination. Nevertheless, if those present chose to help him win election, he would continue to work for the interests of Georgetown.

The Greenbackers then nominated from the first ward, Dr. W. J. Eagles; second ward, Robert O. Old; and third ward, Charles J. Yates. Their slate was complete.

Local Republicans convened the afternoon of March 16, 1878. Judge L. H. Shepard chaired the gathering. They began by holding an informal ballot to select which of several candidates would be nominated for police judge. Next, they held a formal ballot from which William Callery emerged the victor. After Callery was introduced, he thanked the group. He stated that if elected he did not know what direction he would take, but he would attempt to fill the office to the best of his ability.

Republican candidates nominated from the three wards were first ward, William Spruance; second ward, Daniel Roberts; and third ward, Thomas Barnes. After the elections the meeting adjourned.[15]

The general election occurred on Monday, April 1, 1878. Everything went off peacefully, "other than the inevitable verbal zeal always provoked by a close political contest." Fillius emerged the winner, with more votes than two years previously. The selectmen chosen were W. Spruance, a Republican, with

128 votes from the first ward; P. McCann, Democrat, with 91 votes from the second ward; and Thomas Barnes, a Republican, with 28 votes in the third ward. A total of 805 of the town's 821 registered voters cast their vote. Both McCann and Spruance had served on the Board of Selectmen before, whereas Barnes was a liberal-minded man "who was certain to become a good councilman."[16]

The following Monday evening, April 8, Fillius addressed his new government after the welcoming and swearing in of new officers. He noted that the cry for economy was "becoming stereotyped, stale and unprofitable in politics," but he urged its necessity. Georgetown was financially stable, but during the next six months tax receipts would be extremely light. Without exercising due care with expenditures, the community would sink into debt. He wanted to avoid that situation at all costs, since Georgetown's residents were already overburdened with taxes. Therefore, the town fathers should not increase them in spite of "demands from a hundred different sources for money." They should refuse such requests or surrender to bankruptcy.

He pointed out that during the past year the town's streets had absorbed much of the council's time and used up much more money than would be required during the ensuing year. To keep expenses down during 1877, the council had bought a horse and cart and had hired a man at a salary of $75 per month to work on the streets, alleys, and bridges under the city fathers' direction. The committee's chairman acted as a street commissioner, without pay. The cost of the entire project had not exceeded $100 per month, except when the city had been forced to hire extra men. Fillius hoped that because the plan had worked with the preceding council, the new one would also adopt it because street expenses had to be scrutinized constantly, or more money would be spent than the city could afford.

Then he turned to the Fire Department, a noble organization deserving the council's utmost consideration—although not all its requests could be granted. Fillius concluded by urging council members not to be overly prodigal on the one hand or too parsimonious on the other.[17]

Once Georgetown dwellers had settled the matter of local officialdom, they turned to affairs of state and national representation—particularly the issue of U.S. senator. The main question was, who would succeed Jerome B. Chaffee.

After Colorado's admission to the Union on August 1, 1876, the legislature chose Chaffee, a banker and mine operator, and Henry M. Teller, a lawyer, to serve as senators. They were sworn and seated on December 4, 1876, with staggered terms of office. Chaffee drew a two-year term that expired March 4, 1879, whereas Teller's four-month term expired March 3, 1877. On December 9, 1876, Teller was reelected to a full six-year term dating from March 4, 1877.[18]

Senator Chaffee, who had owned a bank in Georgetown during the late 1860s and early 1870s, became ill during 1877, probably as a result of the strain he had undergone while securing Colorado's statehood.[19] Although a report on New Year's Day 1878 stated that his health was "greatly improved" thanks to a visit to Hot Springs, Arkansas, friends began to fear his sickness might be terminal.[20] Colorado Republicans realized Chaffee was too infirm to direct the 1878 campaign, but they also acknowledged his continuing leadership of the party—something he had done since the 1860s. His temporary replacement as chair was W. H. Pierce, to whom Chaffee wrote a letter on May 30, 1878, from New York stating that his physical condition forced him to decline to stand for reelection to the Senate and necessitated that he withdraw from active participation in Colorado politics.[21]

Colorado's Republicans tried to get Chaffee to change his mind, but he remained adamant in his decision not to become involved in the campaign. Others were anxious to serve in the Senate, including three former territorial governors—John Evans, John Routt, and Samuel Elbert. Other hopefuls included George Chilcott, a former territorial delegate; Judge Thomas M. Bowen from Del Norte; and Professor Nathaniel P. Hill, manager of the Boston and Colorado Smelting Company. Hill was not considered a strong candidate, as he was building a new smelter complex at Argo that was said to be much more beneficial to both Colorado and his own interests than his running for office would have been.[22]

Nevertheless, the Republicans settled on Hill, who wrote to Chaffee to secure his cooperation and endorsement before committing himself. Chaffee's reply encouraged the smelterman to seek the office, although the former's poor health would prevent him from being in Colorado during the campaign to aid Hill.[23] Chaffee indicated that he disliked the position of senator and desired to remove himself from politics—factors that had also led to his decision. Later, actions by Chaffee supporters led Hill to publish Chaffee's two letters, a move that angered the ailing senator and led him to make several derogatory statements about Hill. All this created bad feelings between the two men. This situation proved to be the start of ongoing dissension within the Colorado Republican Party.[24]

Before the bad feelings developed between the two men, Chaffee's assurances led Hill to become a Republican candidate and to contribute large sums of money to the campaign. Most of Chaffee's erstwhile supporters then threw their support to Hill, and many of the professor's personal friends insisted that no Republican had as good a chance as Hill to win the Senate seat. In fact, Chaffee had advised his friends to support Hill. In early September Chaffee, who was in Denver at the time, restated his determination not to run: "Personally, I shall not enter politics, that is, so far as running for any office is concerned. I am through now, and want to quit." His statement was reinforced in late November when a report surfaced that Senator Chaffee was "quite ill in Denver."[25]

Thus Hill remained the party's main choice, although Thomas M. Bowen, supported by rebellious factions in southern Colorado, was a strong candidate.

In addition, there were hints that Georgetown's William A. Hamill might take over if an impasse occurred between the main candidates. Nevertheless, later reports showed that Hamill had come out in support of Hill and had worked for him. Besides, Hamill was running the campaign for the Republicans.[26]

Unfortunately, some of Chaffee's more avid supporters persisted in trying to convince him to run. Their stance not only placed the senator in an awkward position but it also divided his friends, who now regarded each other with "deep hostility." Many had already pledged themselves to Hill and were honor bound to support him. On January 1, 1879, the legislature convened, and Hamill, chairman of the State Central Committee, stated that Chaffee still stood by his letter of declination. Hamill could speak with some authority concerning Chaffee's stance because he and the senator were closely involved in mining and political affairs during the 1870s.[27] Nevertheless, Chaffee's health had improved, which led him to state that he would run for senator but only if the party could not agree on any other candidate—a situation so remote that it was unthinkable. This announcement was taken as a repudiation of his previous stance and an indication that he was entering the race against Hill. Such evidence of bad faith did not support his cause.[28]

The Republicans nominated Hill on the fourth ballot thanks to his powerful friends, the greater part of whom had originally been Chaffee's supporters. Nathaniel Hill became Colorado's senator, but he was embittered by Chaffee's actions—which were more the fault of his supporters—and did everything in his power to harass the former senator's friends, including Senator Henry M. Teller in Washington, D.C. The bad blood between Chaffee and Hill caused a schism that greatly affected the Republican Party and nearly caused its demise.[29]

The railway era brought new faith and prosperity to Georgetown. Vast quantities of ores—which a few years before would have been valueless because of the great cost of transportation—now were shipped to Black Hawk, Denver, Golden, Omaha, and overseas. The silver belt from which the shipments originated measured about twelve miles wide by twenty miles long.

Silver mining was Georgetown's chief source of income, but when it came to reporting the annual production of individual mining properties, owners displayed a "considerable degree of reluctance." Many remained silent, which led the Colorado Miner to "guess at the figures." Although a mine's profits and losses were private matters, such statistics were important, not to mention interesting. The world needed to know something about Colorado's gold and silver yield, just as it wanted to learn the numbers of bushels of wheat produced in certain states. The Miner, however, was forced to report information about the work done in 1878 with less confidence than it would have had if the figures had been more complete and precise. Where the editors guessed, they underestimated rather than overestimated the monetary figures. Actually, they used very few numbers, except for those involving the recently discovered Dunderberg—Sherman Mountain's largest producer—whose yield they estimated at between $250,000 and $260,000. Later, Frank Fossett gave its production at $255,000.[30] Other mines were mentioned, such as the Frostberg, whose profitable output since May that year had occurred for several reasons. The Consolidated Phoenix and Coldstream, the Mendota, the Backbone, and the Cap Wells were being worked and were producing at different scales.[31]

On Brown Mountain the Terrible Mine was the most extensively developed property. Its owner, the Colorado United Mining Company, possessed adjacent properties such as the Silver Ore, the Brown, and the Chelsea Beach—all included in the *Colorado Miner*'s report. No production figures were given, although observers had noted a body of ore several hundred feet thick. Fossett estimated the mines' combined 1878 output to be $150,000.[32] A large number of lessees had operated the various lodes; around 200 men were probably laboring in the mines and on the surface. About 20 men were employed on the Shively, which regularly produced ore above average in quality. Other, less important lodes, such as the Atlantic and the Owasco, helped increase the county's output. On Hanna Mountain, west of Silver Plume, eight lessees had been taking out quantities of ore from the Silver Cloud, an old producer.

Republican Mountain had regularly supplied Georgetown's mills with ore. The Pay Rock gave about 50 men steady work as they improved and developed the lode with good results, totaling about $60,000 in 1878.[33] The Vulcan was profitable, and every foot of development increased its value. Exploratory work continued on the Loretta during 1879, even though it had been worked steadily but had not shown a profit. About 50 lessees had toiled all year on the Baxter, whose very promising lower workings had produced about $60,000 that year.[34]

Although the Pelican was practically dormant pending legal decisions, some work had been done on the Dives, particularly on a shaft. Nevertheless, Fossett estimated the 1878 output for each of these mines to be $60,000. A group of lessees tunneling to the Corry City Lode had intersected several new veins that looked good and were being developed. Its general outlook was favorable.

Leavenworth Mountain had produced steadily all year, and its ore was likely higher grade than that taken from any other section. The Equator had been worked throughout 1878 and had seen some very rich strikes, with the ores carrying unusually large amounts of gray copper and ruby silver and producing between $75,000 and $100,000. It continued to have good ore and employed a large number of men. The owners of the Kirtley had recovered very good ore from it, garnering $60,000 or more in 1878. Five or six groups of lessees were hard at work on the Tilden, and although results had not been uniform, exploitation had generally been profitable. The O.K. had not made as much money as it had in 1877, and the Colorado Central had been worked steadily but not extensively. Much drifting had been performed on Marshall Tunnel No. 6, but the results were unknown. The Ocean Wave produced good ore, as its main drift had been cleaned out and retimbered during the preceding nine months. The Argentine, Brannigan, Broadway, Indigo, Alpha, Gilpin,

Welch, and other lodes had been worked continually, with varying but not extraordinary results.

During 1878 some development had been done on the Griffith, which had been idle for several years. Its operators had abandoned it once again, although the *Miner* believed it would prove to be profitable at some point in the future. Elsewhere on Griffith Mountain a tunnel had been undertaken to cut the Comet Lode at a great depth, but there had been no recent news of its progress.[35]

Saxon Mountain had not been as active as formerly, although its lodes had produced quite a bit of mineral. These lodes included the Pickwick, which had some of the richest ore in the county. The Margaret and Sequel lodes were still producing, but the Saxon and the Saxon Extension—although worked by a small group of men—had not reported any results. Most other lodes on Saxon Mountain were in different stages of development.

Democrat Mountain's mines were probably as active as they had been in 1877. Developers had started a shaft on the lower level of the Fred Rogers, to open new ground. On the Little Emma, stopes opening into the main shaft usually carried a fair-sized vein of ore—enough of which was being taken out to encourage the lessees to undertake further exploration. Three parties of lessees were developing different parts of the Junction Mine, whereas workers were doing limited work on the Lucky-Hesperus. The Silver Glance was dormant. Lessees were sinking a shaft on the Silver Cloud and uncovering an increasingly profitable body of ore. In addition, they were tunneling to cut the lode at a depth of 350 feet and were preparing to start another tunnel. The Astor Mine on top of Democrat Mountain had been worked actively during the preceding six months, with satisfactory results. A new wagon road facilitated ore shipments and rendered hitherto worthless ore valuable.[36]

As residents devoted themselves to various economic and political issues during 1878, Georgetown experienced gradual but steady growth. Such growth was proportional to the success of the mining industry, and things looked good at the outset of 1878.

The *Miner* boasted that the Silver Queen was Colorado's second-largest town in population (a boast that certainly could be contested). The tremendous business done by its post office and bookstores illustrated clearly the "character of the population that [had] made the town so celebrated." In addition, its religious instruction and educational facilities were as good, perhaps, as any that could be found in any community in the state. Now that Georgetown had railroad and telegraph communication with the rest of the world—along with its inexhaustible silver mines, excellent public schools, and churches—the town's prosperity was "assured for all time to come."[37]

The Silver Queen endured some problems in 1878, however. The town fathers must have experienced qualms about Leadville, the newest silver mining camp in Colorado, which began its rise to prominence during 1877. During that year many Georgetown residents began moving to the new community to establish themselves. One of these opportunists was David T. Griffith—an original discoverer of the lode that led to the founding of Georgetown—who set up a law practice and began acquiring mining claims in the blossoming camp.[38]

Georgetown's residents had other concerns as well. For instance, the Silver Plume baseball team, named the "Star of the West," defeated the Georgetown Baseball Club so soundly that a handpicked team of Georgetown players challenged them. Alas, the second Georgetown team also lost to Silver Plume by a score of 17 to 16 after a very exciting game.[39]

Then there was the matter of Georgetown children who were disturbing post office employees by continually asking for mail. Many times the youngsters came from families with several children, each of whom called at the post office at different times. Postal workers had to search through letters and papers in response to each plea, which often came in the morning, at noon, and in the evening. There was one daily mail delivery, and when several demands were made throughout the day, it became trying and stretched postal workers' "Christian patience."

One little girl who inquired for mail in the morning and then again in the afternoon was informed both times that there was no mail for her family. After the second time she replied that she knew they had no mail because she had asked earlier.

Children also gathered at the post office throughout the day, passing the time by "knocking off hats, pulling and hauling, scuffling, dancing, scratching the walls and boxes, [and] breaking out window glass," together with other childish pranks. Alex Cree, the postmaster, informed parents that the limited number of postal employees could not spend their time gratifying childish curiosity; nor could they continue to put up with juvenile noise and confusion. He requested parental assistance.[40]

General commerce during 1877 had been fair. Available figures for net sales showed that sales of groceries and provisions totaled $600,000; dry goods and clothing, $325,000 to $350,000; and hardware and mining appliances generally, over $200,000. Individual deposits at the Bank of Clear Creek totaled $2,541,307.89. Although the post office would not release its money order business figures, a safe estimate would be over $100,000 issued.

General F. J. Marshall, agent of the Kansas Pacific Railway, stated that during 1877 his office had shipped 362 carloads of ore, equaling 3,982 tons, at an estimated currency value of $876,000. Later, Frank Fossett gave the exact figure as $800,000.

Coach travel during 1877 had been good, with full loads of passengers until August when the railroad arrived and began transporting larger numbers of travelers. The registers of several hotels showed these numbers of arrivals in 1877: Barton House, 5,550; Yates House, 4,136; and Ennis House, 2,800; while other hotels also accommodated a great many tourists.

The Clear Creek Water Company laid 2,500 feet of main pipes and installed four new hydrants in 1877, resulting in a total of 8,900 feet of mains and sixteen hydrants.[41]

The Colorado Central completed its extension to Georgetown on August 1, 1877, with the first passenger train arriving at 12:45 P.M. The freight depot did not open for several days. Between August 13 and December 31, 5,057,035 pounds of freight were shipped from the Georgetown station, and 8,549,155 pounds were received. The shipments were nearly all silver and lead ores; the receipts were coal, hay, merchandise of all kinds, and feed and provisions.

During 1877 the value of improvements had not reached the customary annual expectations; still, quite a few buildings had been erected—a creditable showing during a year that was more financially depressed in Colorado than any other during the half decade since 1873, when the nation as a whole entered a depression that was nearing its end.[42]

At the end of May 1878, Georgetown was called the "handsomest city in the Rockies—none could be more eligibly located and few have made better use of natural charms." Every one of its residents was directly or indirectly interested in mining. Visitors saw silver bricks displayed at banks and mills and walked around enormous masses of silver ore at hotels and on street corners. Throughout the town, people observed windows and counters lined with silver specimens. Georgetown's inhabitants were very upbeat and enthusiastic about mining and possessed every confidence in their town.

Visitors found Georgetown without rival for multiplicity, beauty, and interesting attractions—not the least of which was Green Lake with its fine stock of fish. The community's accommodations were the best in the mountains.[43]

Georgetown's residents prided themselves on their floriculture. In spite of the high altitude, flowers seemed to thrive: "In almost every house one sees a variety of choice plants and flowers which load the air with their perfume." It was the ladies of the town who devoted time and attention to growing flowers to beautify their homes.

Mrs. F. M. Taylor had a conservatory filled with creeping vines, foliage plants, and fuchsias, among others. Mrs. Dr. Weiser and Mrs. Prof. Weiser grew geraniums of various kinds, fuchsias, heliotropes, and a Chinese primrose, among other plants, many of which were in full bloom when a visiting penman paid a call. Mrs. Prof. Weiser was particularly proud of a *Canna* lily, the only one in Georgetown. Mrs. Kempton cultivated houseplants, plus a variety of old favorites planted in her garden. Mrs. George Sites gave the reporter from the *Miner* a buttonhole bouquet for his adornment. Mrs. A. Medley had one of the finest floral windows to be found. Mrs. Randall was particularly adept at transplanting wildflowers.

Several men also cultivated flowers. Among them was the Rev. John Wilson, pastor of the Methodist church, who possessed a first-rate collection of flowers and had recently constructed a bay window to give his plants every advantage. Dr. R. B. Weiser, a dentist, kept a snapdragon, an apple geranium, a heliotrope, and many other plants in his office. Mr. C. A. Martine, a mill man and mining entrepreneur, had a collection of flowers but was particularly fond of his fuchsias.[44]

Georgetown possessed other accoutrements that went with the good life. For example, in October 1878 Louis Dupuy informed Georgetown residents that he had remodeled his kitchen and was "fully prepared to satisfy the most fastidious taste." He reminded readers that oysters were a specialty in his first-class restaurant. A. Hirsch advised Georgetown dwellers that Anheuser beer had taken first prize at the Paris Exhibition and that he had just received an entire carload of the beverage. Gus Buechner at the "Mint" sold bottles of Budweiser at $3.50 per dozen and $10.00 by the case. Those who desired something stronger could visit the Congress Billiard Hall to sample pure Kentucky whiskies, imported Holland gin, and imported French brandy for only fifteen cents a drink. As the holiday season began, Hirsch gave his customers the glad tidings that they did not have to go to Denver to replenish their liquor cabinets, for he had "a full and complete stock, including Scotch and Irish whiskies." All these beverages were first class and could be purchased at wholesale or retail prices.[45]

Finally, one tobacco shop proprietor told prospective customers:

IT HAS BEEN DECIDED

By the best physicians in this country and Europe, that tobacco is an antiseptic—that is beneficial to health and promotive of long life. You can find the best of cigars, domestic and imported, also a fine line of meerschaum goods—pipes and cigar holders—suitable for Christmas presents, at Fred Aldinger's "Senate."[46]

Those in search of a house would have been interested in this advertisement:

FOR SALE

The property of George Schutz, on the east side of Argentine street between 6th and 7th. The house is new, containing four rooms and [a] kitchen. All plastered. A never failing well with pump, stable, woodshed, all convenient. For further particulars call on JNO. S. FISHER, Agt.[47]

Georgetown's population also enjoyed art and was notified through the columns of the *Miner* about a fine exhibit of artwork at Murphy Hall that would be shown on December 21, 1878. Admission was ten cents. Prospective spectators were cautioned not to touch the paintings or the statuary.[48]

The Silver Queen's subjects actively participated in Fourth of July and Thanksgiving celebrations. Readers of the *Colorado Miner* learned at the end of June that the Emerald Rifles, Hope Hose Company No. 1, and a number of patriotic citizens were arranging appropriate ceremonies to make the Fourth a memorable one. Over $300 had been subscribed, and another $100 to $200 would probably be raised before the celebration.

The organizers agreed to offer prizes for the fireman's race: "1st Prize' $75.00; 2nd Prize' $50.00; 3rd Prize' $25.00." The Fireworks Committee had raised $100 and had spent it on an appropriate supply of noisemakers. Another committee was appointed to invite the mayor, the council, and the city to assist with the celebration. This was a wise measure, since "the City [had] declined making any appropriation." Another committee was to "invite thirty-eight young ladies to participate in the celebration by representing the States of the Union in the procession."

Festivities would begin at sunrise with a salute from the local cannon, "Iron-Clad Dick," which would then be turned over to the Hope Hose Company for the rest of the day. A parade would begin promptly at 9:30 A.M. so there would be plenty of time for the afternoon races. The procession would end at a speaker's platform on Taos Street where the Honorable Thomas J. Cantlon would deliver an oration, followed by Frank R. Carpenter, Esq., reading the Declaration of Independence. That evening the grand ball of the Hope Hose Company and the Emerald Rifles would go on at Cushman's Opera House. In addition to the festivities in Georgetown, a picnic and boat races also took place at Green Lake, which ended at 5:00 P.M. thanks to a shower.

Georgetown's Fourth of July celebration went off quietly. Only six or seven arrests were made as a result of fights, drunkenness, and someone firing a pistol in the street. Judge Fillius "made due allowance for its being July 4, and was merciful. The consequence was five fines, the balance was charged to the light air and tarantula juice."[49]

Thanksgiving, more of a family holiday, was even quieter. President Hayes proclaimed Thursday, November 28, Thanksgiving Day. Hope Hose Company No. 1 made arrangements for its fifth annual ball at Cushman's Opera House the evening of November 28. The price of admission for gentlemen with ladies was $2.50. The ball appeared to be the only public observance of the feast day and was proclaimed both a financial and a social success.[50]

In 1878 everyone eagerly awaited the Christmas holidays. Numerous miners came out of the mountains, and "family men devoted considerable of their spare change or strained their credit in purchasing gifts for their [wives] and children." Others partook of stimulating grain extracts or the amber beverage jolly old King Gambrinus had once quaffed, but few overindulged. (Gambrinus was a legendary Flemish king who supposedly invented beer.)

On Christmas Eve several happy household gatherings were held, featuring Christmas trees and delighting little ones. The Methodist church held services conducted by its Sunday School scholars and others. Grace Episcopal Church attracted multitudinous worshippers who witnessed the distribution of a large number of presents from the Christmas tree and heard excellent music. The Catholics postponed their midnight Mass, along with its music, until 5:00

A.M. so there would be fewer outsiders to interfere with the devotions of the faithful.

Light snow fell that cold Christmas Eve, "giving promise of an old fashioned Christmas in the States' and reminding residents of a heartening old saying: "A Christmas white means a graveyard lean."

In spite of early climatological indications, Christmas was clear and cold, which gave Georgetown's populace the excuse to fill its streets. Saloons provided customers with lunches made up chiefly of turkey; while one served arrack punch, another dispensed eggnog, and a third tacked "Merry Christmas" on a box of free cigars. (Arrack was a potent Far Eastern alcoholic drink that tasted like rum but was distilled from fermented juice of the coconut palm or from a fermented mash of rice and molasses.)

Christmas afternoon saw dinners and social affairs throughout the community. Guests at the home of James R. Morris, Esq., ate a sumptuous repast and then were treated to music and magic lantern views. Captain and Mrs. Phillips celebrated their twentieth wedding anniversary, at which they were presented with "a magnificent China tea set."

That evening Georgetown Fire Company No. 1 hosted a ball at Cushman's Opera House. It was a brilliant financial and social success. The tasteful decorations included evergreen wreaths encircling each column in the hall, and "Merry Christmas" in green letters hung over the bandstand. Gordon's Band furnished the music, to which an estimated 200 people danced.[51]

New Year's Day was "tame and thin blooded" compared to Christmas. The town was quiet, although a number of businessmen formed a group and made calls on local residents. At Judge Coulter's residence Miss Stella received friends, "both young and old," and, accompanied by Misses Mollie Shadhalt and Georgie Kembley, entertained them by singing several songs. Early that evening Conrad Hanson, a young bachelor, held a party in his apartments.

At about 8:30 P.M. over seventy-five ladies and gentlemen crowded aboard a train to attend the firemen's ball given by Idaho Hook and Ladder Company No. 1. The fest took place in the large dining room of Beebee House in Idaho Springs, where guests danced to the music of Gordon's Band until 5:00 A.M. when the train started back to Georgetown.[52]

Life was good in the Silver Queen during 1878, as the town continued its steady and permanent growth.[53] As Georgetown entered its twentieth year, it had surmounted its unstable years as a mining camp and developed into a staid community. Its political and social institutions functioned well, the populace was hardworking and sober, and residents appeared ready to join the middle class that was starting to emerge throughout the United States. Georgetown's future appeared bright thanks to the Bland-Allison Act, which assured the miners that silver—although not completely restored to its former status on coinage lists—would at least be purchased in sufficient quantities by the federal government to ensure a market for the silver produced by Georgetown's mines. Its ore production was backed by the community's climate, scenic attractions, and fishing—all of which enticed not only tourists but persons with mining interests who sought the solace of a beautiful area in which they might luxuriate while investing and working. In an era that long preceded air conditioning, Georgetown's mild climate was a haven for Americans attempting to escape the summer heat.

The one exception to the rosy picture that seemed poised to take the community into the 1880s was the new mining camp of Leadville, whose vast silver resources threatened the dominance of Colorado's premier silver camp. But few of the revelers returning to Georgetown that cold, crisp morning of January 1, 1879, gave much thought to that possibility.

NOTES

1. *Rocky Mountain News,* January 1, 1878, 2.
2. *Colorado Miner,* February 16, 1878, 1.
3. Allen Weinstein, *Prelude to Populism: Origins of the Silver Issue, 1867–1878* (New Haven: Yale University Press, 1970), 239–240.
4. *Rocky Mountain News,* January 10, 1878, 1.
5. *Colorado Miner,* February 9, 1878, 2.
6. *Colorado Miner,* February 23, 1878, 3.
7. *Colorado Miner,* March 2, 1878, 2.
8. Ibid.
9. *Colorado Miner,* February 23, 1878, 1.
10. *Colorado Miner,* March 2, 1878, 2.
11. *Colorado Miner,* March 9, 1878, 1.
12. *Colorado Miner,* April 13, 1878, 2–3.
13. *Colorado Miner,* April 6, 1878, 3.
14. *Colorado Miner,* March 16, 1878, 3.
15. *Colorado Miner,* March 23, 1878, 3.
16. *Colorado Miner,* April 6, 1878, 3.
17. *Colorado Miner,* April 13, 1878, 3; April 20, 1878, 3.
18. Frank Hall, *History of the State of Colorado* (Chicago: Blakely, 1889), II: 360.
19. Ibid., 488; See also Robert S. Pulcipher, ed., *The Pioneer Western Bank—First of Denver: 1860–1980* (Denver: First Interstate Bank of Denver and State Historical Society of Colorado, 1984), 40–41.
20. *Rocky Mountain News,* January 1, 1878, 1; Hall, *Colorado,* II: 489.
21. Hall, *Colorado,* II: 489.
22. *Colorado Miner,* June 15, 1878, 2.
23. Hall, *Colorado,* II: 489–490.
24. R. G. Dill, *The Political Campaigns of Colorado, With Complete Tabulated Statements of the Official Vote* (Denver: Arapahoe, 1895), 41–42.
25. Hall, *Colorado,* II: 490; *Colorado Miner,* July 13, 1878, 2; quote in September 7, 1878, 2; November 23, 1878, 3.
26. *Colorado Miner,* December 14, 1878, 2; January 11, 1879, 2; Dill, *Political Campaigns,* 36–39.
27. Christine Bradley, *William A. Hamill: The Gentleman From Clear Creek* (Fort Collins: Colorado State University Cooperative Extension Service, Historical Bulletin no. 2, 1977), 28.

28. Hall, *Colorado,* II: 491; Charles S. Thomas, "Fifty Years of Political History," in James H. Baker and LeRoy R. Hafen, eds., *History of Colorado,* 5 vols. (Denver: Linderman, 1927), III: 909.

29. Hall, *Colorado,* II: 492; Thomas, "Fifty Years," III: 909.

30. *Colorado Miner,* May 25, 1878, 1; January 11, 1879, 3; Frank Fossett, *Colorado; Its Gold and Silver Mines, Farms and Stock Ranges, and Health and Pleasure Resorts* (New York: C. G. Crawford, 1879), 364, 396.

31. *Colorado Miner,* January 11, 1879, 3.

32. Fossett, *Colorado,* 364.

33. Ibid., 364, 394.

34. Ibid., 364.

35. Ibid., 364, 391; *Colorado Miner,* January 11, 1879.

36. *Colorado Miner,* January 11, 1879, 3.

37. *Colorado Miner,* January 12, 1878, 4.

38. John Willard Horner, *Silver Town* (Caldwell, Idaho: Caxton, 1950), 273; *Colorado Miner,* October 26, 1878, 3.

39. *Colorado Miner,* June 15, 1878, 3.

40. *Colorado Miner,* March 9, 1878, 2.

41. Ibid; Fossett, *Colorado,* 362.

42. *Colorado Miner,* January 12, 1878, 4; Fossett, *Colorado,* 362.

43. *Colorado Miner,* May 25, 1878, 1.

44. *Colorado Miner,* August 3, 1878, 3, 6 (quote on p. 3); O. L. Baskin, *History of Boulder and Clear Creek Valleys, Colorado* (Evansville, Ind.: Unigraphic, 1971 [1880]), 290.

45. *Colorado Miner,* October 26, 1878, 3; December 21, 1878, 3.

46. *Colorado Miner,* December 21, 1878, 3.

47. *Colorado Miner,* December 14, 1878, 3.

48. Ibid., 2.

49. *Colorado Miner,* June 29, 1878, 2; July 6, 1878, 2.

50. *Colorado Miner,* November 9, 1878, 3; January 11, 1879, 1.

51. *Colorado Miner,* December 28, 1878, 3.

52. *Colorado Miner,* January 4, 1879, 2.

53. *Colorado Miner,* January 11, 1879, 2.

— What to Do, What to Do? —

Life in a mountain valley, with limited sunlight and wind swirling around the streets, could be gloomy and depressing. Today we turn to satellite dishes, radio, and the Internet. Nineteenth-century Georgetown lacked the technology we depend on; however, residents were never at a loss for entertainment.

Although not the first balloon ascension in Georgetown, this is probably the first photograph of one, taken by Harry H. Buckwalter in 1901 during a failed effort to take aerial photographs of the Georgetown Loop. Courtesy, Colorado Historical Society, CHS-B212.

Who could forget "Le Celebre Leotard" leaping from one side of Sixth Street to the other, followed by a grand trapeze performance in which "[h]e hung by his toes and knees, and then swung on one arm, and caught on his ear, and then balanced on one eye-lash while playing Yankee Doodle on a guitar with his two hands, and did besides a number of other things too numerous to mention" (*Colorado Miner* [Georgetown], June 26, 1873).

Hot air balloons came through town on several occasions: "The aerial ship that left Denver yesterday morning arrived in Georgetown a little before 11 A.M., and was greeted by a large number of our citizens headed by the Silver Queen band playing 'Up in a Balloon, Boys.' The altitudinous voyagers took dinner at the Barton and left for Kokomo, via. the 'high line,' at 1 o'clock" (*Colorado Miner,* April 2, 1881).

Dancing Russian bears, Buffalo Bill and his Wild West show, several circuses, and the far-famed Cardiff Giant all made their way to the Silver Queen. The local paper described the Cardiff Giant: "This curiosity should be seen by all. The length is ten feet four and one-half inches, and [it] weighs 2,290 pounds. It is dead and perfectly harmless" (*Colorado Miner,* July 5, 1879). What more could one hope to see?

E.H.N. Patterson, "Sniktau," with full beard. Courtesy, Colorado Historical Society, F-51.

————— Sniktau —————

E.H.N. Patterson brought his quick wit and keen obser-vances to the pages of Georgetown's *Colorado Miner* between 1873 and his death in 1880. A veteran of the mining camps in California (where he worked on the editorial staff of Sacramento's *Placer Times*) and early Colorado (1859–1860, submitting articles to the *Rocky Mountain News* [Denver], *Western Mountaineer* [Golden], and the *Golden Transcript*), interspersed with years as a writer for the Washington [D.C.] *Jacksonian* and writer/editor of the Oquawka [Illinois] *Spectator,* he gave the Georgetown paper several years of a journalistic professionalism not always seen in mining camps.

Patterson first headed west in 1849 when his plans to publish a journal with Edgar Allen Poe collapsed after the poet's death. The young man loved the freedom of undeveloped lands and cultures, combining humor and insight with his infectious love of the West.

Three generations of Pattersons (including E.H.N.) were editors of the Oquawka *Spectator,* yet the lure of gold and silver kept calling him. Writing under the pen name "Sniktau," he provided readers across the country with firsthand accounts of the development of the West. His writings featured a broad brush of nationalities, colors, and walks of life, providing an in-depth look at Georgetown and the surrounding communities.

When Sniktau's writing was combined with the pen of the colorful Stephen Decatur or the scholarly Reuben Weiser, the pages of the *Miner* read like a good novel and provided rich fodder for research for generations to come.

THE EARLY SILVER MINES OF WESTERN CLEAR CREEK County encompassed a wide area stretching from the Argentine districts straddling Mount McClellan, running through Bakerville, Brownville, Silver Plume, Silver Dale, and Georgetown, then extending north and east to the small towns of Alvarado, Swansea, Silver Creek, Red Elephant, and Lawson. Georgetown became the commercial center—the supply point for most of these early ventures—but the mines and mills spread across and beneath miles of mountainsides.

The first mills were built close to the mines. Optimistic mine owners hauled tons of equipment to high altitudes to avoid hauling tons of rock down the hillsides. Dibben's Smelter in East Argentine and the Baker Company's mill in Bakerville stood as testament to the stubborn determination of their owners and the strong backs of burros. Soon, others located mill sites in the valley floor of Georgetown, avoiding the steep climb up hillsides to the south and west and taking advantage of the ready supply of water power. A group of British investors built a large smelter at a new town site at the junction of the Middle and South forks of Clear Creek, which they named "Swansea," hoping to share the success and reputation of the town of Swansea, Wales.

Mills had a habit of burning to the ground; the tremendous heat required to process ore often sparked fires in the chimneys or on the rooftops. As a result, the early mills led short lives; however, many were captured by early photographers and are presented here, along with other facets of early mining and milling around Georgetown.

Mining and Milling

Above: *The Wilson and Cass Mill (built 1869) at Plummer's Dam—now the head of Georgetown Lake and Locke Dam. The mill sits in the general vicinity of the town's sewer plant. Courtesy, Harold Frost Collection. Below: Frank Dibbens's mill in East Argentine, also known as the International Smelter, built in 1868–1869, was a popular stop for tourists riding the Argentine Central Railroad in the early twentieth century. Courtesy, Denver Public Library, Western History Collection, X-1129.*

Above: *The Baker Company Reduction Works opened for business in September 1869, then burned to the ground on August 5, 1871. Photographer William Chamberlain managed to catch the works in full production. Courtesy, Denver Public Library, Western History Collection, X-19313.* Below: *The small town of Bakerville grew up at the junction of Stevens (Quail) Creek and the South Fork of Clear Creek. This early photo shows the reduction works, stable, and boardinghouse ca. 1870. Courtesy, Denver Public Library, Western History Collection, Chamberlain photo, X-19317.*

Two means of transportation are seen in one photo: on the left is the "Graymont sluice" used to transport lumber from the mountains around Graymont (just east of Bakerville) to sawmills in Silver Plume. On the right is the more efficient but more expensive railroad. View looking west toward Loveland Pass. Courtesy, Harold Frost Collection, William Henry Jackson photo.

Two views of the Georgetown smelter (also known as the Georgetown Silver Works, later the Hall Mill), ca. 1870, built by Jerome Chaffee, John Stryker, Eben Smith, and John T. Herrick, run by water piped from a dam at "Devil's Gate" in Clear Creek through a wooden aqueduct prone to leakage. Above: View to the southeast showing the back of the original Watson house. Courtesy, Colorado Historical Society, Eggleston Bros. photo, 84.192.1640. Below: Photo looking northwest showing brick chimney on Chimney Rock. Courtesy, Denver Public Library, Western History Collection, Hull photo, Z-5767.

The label on the back of the upper photograph says it depicts armed miners during the Dives-Pelican feud (mines above Silver Plume), with the rifles and handguns as proof. The lower photo is of the same group, with more picks and shovels than guns—perhaps a kinder, gentler photo to send home to their families? Courtesy, Denver Public Library, Western History Collection, Randall Collection, F-5163 (above), F-5161 (below).

Above: *Brownville, the town located at the foot of Brown Gulch, started to develop in 1868 as a home to many early miners. Courtesy, Denver Public Library, Western History Collection, X-7235.* Left: *Transporting ore from mines high above the valley required a steady hand and an experienced team. It is no wonder the mine owners invested in tunnels as a safer alternative. 7:30 road above Silver Plume. Courtesy, Tutt Library, Colorado College, Colorado Room Photo Files PP 85-31s.*

Above: *Just to the east of Brownville, Silver Plume soon developed in a better location, less prone to mud and snowslides. This photo is one of the earliest known views of Silver Plume, ca. 1870. Courtesy, Denver Public Library, Western History Collection, X-13571. Below: "Freighting Supplies to Argentine District," taken on the road to East Argentine from above the Marshall Tunnel in the Silver Dale area. Courtesy, Historic Georgetown, Inc., Dagleish photo.*

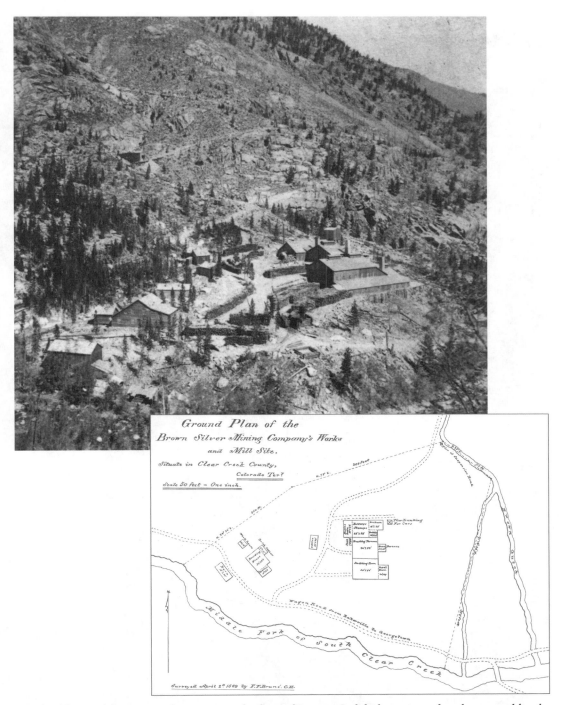

The buildings of the Brown Company at the foot of Brown Gulch, later owned and operated by the Colorado Terrible Lode Mining Company, Ltd. The plat filed with Clear Creek County identifies the buildings in the photo above (Book R, 493, Clear Creek County Archives, Georgetown). Courtesy, Denver Public Library, Western History Collection, X-7236.

This 1868 photo, taken by Arundel Hull, shows Stewart's first large mill (he built a small mill in 1865 on the "left-hand fork" below the town's water plant), built in 1868 between the 22nd Street bridge and I-70 mile marker 228, probably on the interstate right-of-way. Stewart built a new mill in 1870 near today's tennis courts. The new building burned to the ground in January 1871, was rebuilt, and burned again in December 1875. Many felt the second fire was an economic blow from which the town never fully recovered. Courtesy, Denver Public Library, Western History Collection, Z-5765.

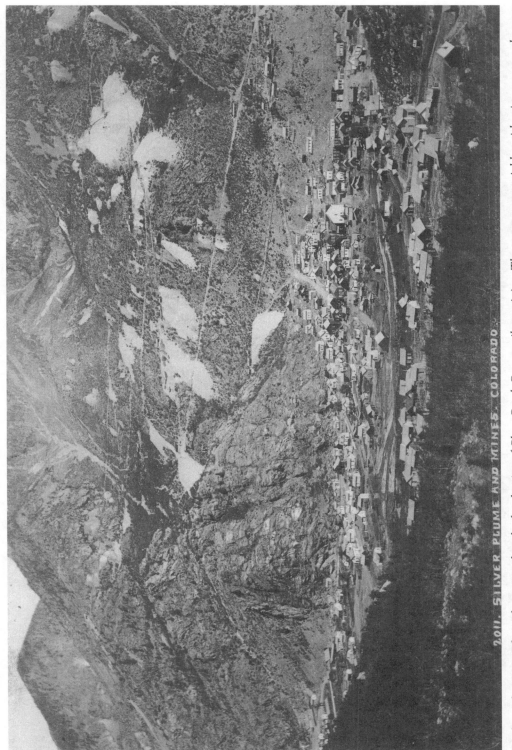

2011. SILVER PLUME AND MINES. COLORADO.

By the early 1880s Silver Plume stood as the focal point of Clear Creek County silver mining. The town grew quickly, with mines scattered across the face of the mountains to the north. Courtesy, Historic Georgetown, Inc., photo donated by James R. Fitzpatrick.

Above: *Payrock Mill at the east end of Silver Plume between I-70 and Clear Creek. Note the terminus of the aerial tram in the upper right-hand corner of the picture. Courtesy, Denver Public Library, Western History Collection, X-61593. Below: The mines of the Colorado Central Company dominated the landscape around Silver Dale, a small town south of Georgetown. Courtesy, Historic Georgetown, Inc.*

Above: *The mill in the foreground is the Judd and Crosby Works, later known as the Farwell Reduction Works. By 1873 the Colorado Miner [Georgetown] reported that between $75,000 and $100,000 had been spent on the mill. The money poured into these operations far outweighed the returns. Courtesy, Harold Frost Collection, Collier photo. Right: These bars of silver were stacked carefully to show eastern and overseas investors that wealth was to be found in the local mines. Courtesy, Denver Public Library, Western History Collection, Joseph Collier photo, X-60034.*

Underground photographs are scarce because of problems with lighting; however, local photographers George Dalgleish and Lachlan McLean captured several shots. Above: An interior shot from the Cosmos Tunnel, Doric Mine, Georgetown; W. J. Lewis Sr., W. J. Lewis Jr., and young Bert Snodgrass. The man to the left is "single-jacking"—holding his own drill bit, pounded by a sledgehammer, to drill a hole for explosives. Courtesy, Denver Public Library, Western History Collection, Alice Morton Mattern Collection, X-61609. Below: Double-jacking required concentration, timing, and trust, as shown in this McLean photo. One miner held the drill steel while the other swung the hammer. Courtesy, Archives, University of Colorado at Boulder Libraries, McLean photo, 351.

Above: *Interior of local stamp mill. Courtesy, Colorado Historical Society, Neal Fisk Collection, 10031980.* Below: *Double-jacking in the East Argentine area, probably the Sidney Tunnel. Courtesy, Colorado Historical Society, George Dalgleish photo, F-1136.*

Above: *Electric power gave the mining industry a jump start in the early 1890s, but the trusty miner's candle is still in use in this photo from the Doric Tunnel, Georgetown (ca. 1895). Courtesy, Denver Public Library, Western History Collection, X-61089.* Left: *Leyner drills worked at great speed, but they could not overcome the onset of hard times in the 1890s. Courtesy, Colorado Historical Society, George Dalgleish photo, 20004577.*

7 Lifestyles of the Sometimes Rich and Not So Famous

NOTICE—Frank S Butler advertises me as having left his bed and board. This is a mistake, as I own the bed and took it with me. MALVINA BUTLER.

—*Colorado Miner,* August 13, 1868, 4

FROM THE EARLIEST DAYS OF GEORGETOWN'S EXISTENCE, THE LADIES OF THE EVENING presented dilemmas for the town's god-fearing folk. Young Malvina Butler, quoted above, made the local papers on a regular basis, as she and her spouse were known to fight often and loudly—much to the amusement of the community. The red light district ran north from Seventh Street a few blocks along Brownell, across the creek from the more respectable part of town. Like most other communities, Georgetown looked the other way as long as the problems associated with the district (drinking, drugs, fighting) did not spill out into the rest of the town.

Malcolm McCrimmon—the local blacksmith who lived with his wife and several young children on Brownell, just south of the district—appeared before the Board of Selectmen in 1881 to complain that things were getting out of hand. The ladies flaunted their wares too close to home (his home at least) in direct violation of town ordinances.[1] The subsequent board discussion indicated a split between those who thought the marshal should enforce the letter of the law and stop the operation of bawdy houses and those who felt the marshal had his hands full "segregating the business from respectable quarters and confining it to Brownell street."[2]

Bell Keys, Mattie Estes, Josie Woods, Malvina Butler, and other madams through the years lived as outcasts within local society; however, they were also businesswomen who dealt with the local merchants on a regular basis. With the exception of occasional petitions from neighbors or groups in town frustrated by the community's seeming acceptance of "loose" behavior on the part of these women and the men who frequented their establishments, the town seemed willing to accept the existence of prostitution as long as the problem behavior and most of the "inmates" stayed on the other side of the creek.

The townspeople stood firm in their desire to make Georgetown a safe place to live—an upstanding community known more for its fine homes, mining opportunities, and tourist attractions than were many of the rowdier towns in the Rocky Mountain West. Law enforcement struggled to stay ahead of the fray in the earliest years—times that saw two hangings. Ed Bainbridge, described as a "half-crazed" mulatto, fond of cards and not particularly fond of white men, was lynched in lower town on April 24, 1867, for the shooting of Jim Martin (who, unbeknownst to the mob, had not died from his injuries);[3] and Robert Schramle was hanged over Albert Selak's pigpen in December 1877 for the murder of Henry Thiede at the slaughterhouse below town earlier that year.[4] Events such as these and the murder of well-respected Jacob Snider by Jackson Bishop, a local ruffian involved in the feud between the Dives and Pelican mines outside Silver Plume, put the town on edge (the murder is discussed in Chapter 5). The locals demanded more and better law enforcement to keep the lawless element at bay.

A county jail was needed, and one was built on Biddle Street in 1873.[5] The county calaboose, however, would not be known for its tight security. Jailbreaks were common (and apparently quite simple to achieve).[6] Rumblings for a new stone or brick jail started almost immediately. After an escape in 1882, the *Miner* pointed out the obvious: "In short, it is simply foolishness to expect to keep men shut up in a wooden building which is little more than a rickety outhouse."[7]

Dollars would not be easy for the county to secure for any of its projects. After several ill-fated attempts to ask voters for funds to build a courthouse, the exasperated commissioners finally decided to purchase the Ohio Bakery building on Argentine Street. The two-story wooden structure, built in 1868 by Commissioner John Fillius and his partner Jerry G. Mahaney, would be modified to provide a central building in which all county officials could have their offices. The county acquired the property in 1874 and proudly moved in to the offices in June 1874. Although they undoubtedly felt this was a temporary home, the officials were pleased nonetheless:

> The New Court House. The workmen have been very busy getting the new court room in readiness for the next term of District Court, which commences next Monday. Mr. Boyer assures us that the court room will be ready by that time.

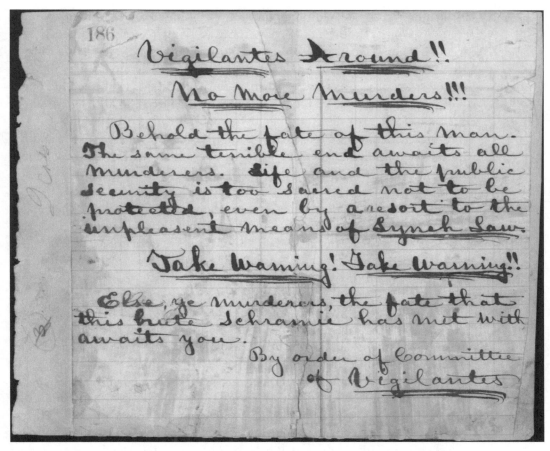

Vigilante notice pinned to the chest of Robert Schramle, hanged by vigilantes in 1877. Courtesy, Denver Public Library, Western History Collection, F-2994.

The lower story will be divided into two apartments, of which the one in front, 24 feet square, will be occupied by the Recorder and Treasurer. This office will be large and roomy, and the large windows will make it very light and pleasant. A good vault in one corner will give ample security to books, papers, and records in case of fire. In the rear of the Recorder's office is the stairway leading to the upper story, which connects with a door on the north side. The court room is 55 feet wide by 24 feet wide, and will be fitted up in good style.

The second story will be occupied as follows: In front is the Probate Court room, 24 feet square; a hall 6 feet wide and 61 feet long runs from this room to the west end of the building, and along the north side will be five jury rooms, each 11 by 18 feet, with two windows in each.[8]

Once the county officers had moved into their new offices, their attention turned again to the need for a new jail. The rest of the county (especially folks in Idaho Springs, still stung by the removal of the county seat in 1868) did not see the need to provide additional funds and continued to defeat bonds for any proposed county building. Even without a bond to help with costs, the commissioners decided to replace the old jail with a new brick building in 1883, to the relief of the sheriff and neighbors on Biddle Street.

The town of Georgetown used some of the materials from the old county jail to build a new, 16 × 18 foot stone jail in the fall of 1883 at the west end of Fourth Street next to the county courthouse, at a total cost of $702.70.[9] The paper nicknamed it the Hotel de Barr, after both Town Marshal Barr and the window decorations.[10] Whereas most serious criminals still went to the county jail, the small town facility could be used for local drunks or vagrants who did not need to be incarcerated with the rougher element.

As the last bricks were being laid on the new county jail, another major project began to wind its way up the valley behind the jail and the residences on Biddle Street. The railroad—then known as the Georgetown, Breckenridge and Leadville extension of the Union Pacific—announced plans to extend the line to the west, hoping to reach the valued Leadville market before other competitors. In January 1882 the *Miner* reported, "The construction of the first mile will cost $30,000 to $35,000, and the right of way will probably cost [$]20 to $25,000."[11] The first mile would be simple, however, compared to the difficulty of building a railroad at a level grade up the narrow, steep canyon between Georgetown and Silver Plume.

To maintain a manageable grade, the Union Pacific engineers designed a "loop" by having the railroad travel up the canyon, double back, and cross over itself, using a bridge 95 feet high and spanning approximately 300 feet that crossed over the creek and the lower level of the track. Although built out of necessity, the high bridge (soon to be known as the Devil's Gate viaduct after a nearby rock formation) would soon become the town's best-known asset and a worldwide tourist attraction: "The Far Famed Georgetown Loop."

Before the bridge was built, however, the four miles of track needed to be carved from the rocky hillsides surrounding the town. Frank Hollingsworth contracted for the first mile of track, housing 50 men at a camp near the Selak Brewery on the east end of Ninth Street and paying $1.75 a day, with $5.00 a week for board. The newspaper noted that 13 black men were hired in February 1882.[12] Shortly thereafter, 50 Italians signed up.[13] Swedish workers took out a subcontract on a portion of the contract running between the Clear Creek and Pelican mills (from the east end of Fourth to Sixth streets).[14] The ethnic diversity was completed with the hiring of a Russian section boss.[15]

Captain H. C. Ballard held the contract for the remaining four miles of track, establishing two camps for 125 men: one up the valley near the Lebanon Mine, and the other near the Boche Brewery on Main Street in Silver Plume.[16] When an effort was made to bring Chinese workers into Silver Plume, local workers rebelled. The Chinese returned to Denver. The *Miner* noted: "There seems to be a mutual impression in Silver Plume and Georgetown that the air is a little too light at this altitude for chinamen, anyhow, and as a sanitary measure they had better seek a more congenial clime."[17]

The work continued, with rocks raining down upon the town as workers attempted to cut through solid granite with high explosives. Dr. J. G. Pohle's barn on the road to Silver Plume disintegrated as a 1,000-pound rock flew over 600 feet across the creek and over his house. At the same time, another 500-pound rock demolished his front fence. The town hired a special policeman, J. A. Love, to monitor the workers and arrest anyone damaging property in town.[18] With a touch of humor the *Miner* proclaimed, "It's bully, but it ain't war"—a phrase attributed to French general Canrobet when he was watching the Charge of the Light Brigade (Crimean War, 1854–1856).[19]

Work continued, however, and in October 1883 a passenger coach traversed the new track up to the start of the loop bridge with a party of dignitaries from the Union Pacific, including New York financier and railroad magnate Jay Gould.[20] The bridge was completed in 1884, and although the Union Pacific lost the race to Leadville, the far-famed Georgetown Loop was a success. Tourists soon heard of the new engineering marvel and traveled from across the country and around the world to ride the narrow-gauge train as it chugged up the valley.

By 1884, travelers to Georgetown came to a town with elegant accommodations, established shops, fenced yards, and carefully tended gardens. The town offered all the comforts of home—indeed, several comforts that might not yet be available elsewhere. By September 1879 Georgetown boasted of telephone connections:

> The Western Union Telegraph Company have [sic] been putting in new telegraph poles and have also added the wires for the Edison Telephone, which is now connected with nearly all the business houses in town, and we are beginning to put on metropolitan airs. During the day we expect to jaw back on our own hook with either Central, Golden or Denver.[21]

By December of that year, however, the editors were already complaining about interruptions as they tried to get things done.[22] Nonetheless, Georgetown's telephone connections pre-dated many towns in other parts of the country and the world, an indication of the value local mining companies and investors placed on scientific innovation.

Visitors to town had their choice of several hotels or boardinghouses, including two particularly fine establishments—the Barton House and the Hotel de Paris. The fine old Barton House, rebuilt in 1872, proudly hosted President Ulysses S. Grant in 1875. The hotel was spiffed up even further with major renovations in 1877:

> A gentlemen's parlor is entered from the office, with an outlook upon Taos and Burrell streets. This is 12 by 16 feet; finished with walnut and curled maple, upholstered with elegant writing desk, walnut-and-leather chairs,

Brussels carpet and venetian blinds. The ladies' parlor, 16 by 24 feet, will be
heated by coal grate, is newly papered, and will be furnished with elegant
carpet, chairs, sofa, center table, piano, marble mantle and pier-glass. The
Billiard Room, 19 by 41 feet, is resplendent with its new paint and elegant
bar. The dining room will lose its yellow tinted walls and be clad in cheerful
French gray. The laundry will be made larger and prepared to do the entire
work in this line for the guests of the house. The lower story of the Hotel
has been refloored with hard maple in 2½ inch strips. On the second floor
three new rooms have been added; a suite of parlor and two rooms prepared
for guests with families; all the rooms have been repainted and kalsomined
in light and cheerful tints; a bath room with hot and cold water has been
added; and two water-closets have been put in for the accommodation of
lady guests.[23]

The stately hostelry, perched on the hilltop in the commercial district, served
visitors from across the world. In 1882, however, with Louis DuPuy's additions
and renovations to his Hotel de Paris, Barton House business began to suffer.
In November 1883 Billy Barton closed Barton House and headed back to
Boston, promising "extensive repairs" over the winter.[24]

The growth of the tourist industry, especially the construction of the
Georgetown Loop, gave the town's best-known Frenchman an excellent reason
to expand the Hotel de Paris in 1882. In 1875 DuPuy started by renting the
Delmonico Bakery building, with plans for a restaurant and boarding rooms.
The *Miner* proclaimed:

Oysters, game, and all the delicacies of the season will be kept, and one of
the best cooks in the Territory will prepare the viands. No bar is attached to
the establishment, but gentlemen who like a glass of ale, porter, French or
Rhenish wine, with their dinners, can be supplied with the best. A fine suite
of furnished rooms will give regular boarders the opportunity of having a
home in the building. We are confident that the enterprise, energy and good
taste displayed by Mr. D. will be rewarded by the patronage of our citizens.[25]

In 1882 Dupuy constructed a brick addition to the west of the former
bakery building, doubling its size. Work continued throughout the year, with
excavation for a wine cellar and construction of fine rooms. The work neared
completion by Christmas:

[The addition is] elegantly furnished and carpeted with Wilton carpets,
heated by steam and in the centre of the old dining hall he proposes to have
a fountain. Louie furnishes the best of everything to a hungry public and
always welcomes his friends with a beaming smile, and it is well worth while
to any one to go through the culinary department and examine the
convenient manner in which every thing is arranged. Louie well deserves his
popularity.[26]

With the brick addition, the Hotel de Paris completed the transition from
a restaurant and boardinghouse to a full-fledged hotel with fine dining. The
engineers and attorneys for the Union Pacific Railway soon gravitated to the

new French inn, as did many of the town's elite. Whether it was the quality of food, the steam heat, or good conversation, Dupuy's hotel soon became the hotel of choice in Georgetown.

The town's commercial district grew rapidly once plans for the extension of the railroad were announced. The tightly packed buildings stood protected from fire by four fire companies with relatively new equipment, piped water, and solid backing from the community. The combination of dedicated volunteers, training, and equipment saved Georgetown from the disastrous fires that beset many Colorado towns.

Georgetown's two hose cart teams recruited the fastest runners, both to get hoses to fires and to guarantee that they would be able to compete with surrounding towns at fire department competitions. Races were held throughout the Front Range. The state tournament, held in Denver in August 1880, brought the best and the fastest to Fourteenth Street to show what they could do. The Alpines, representing Alpine Hose Company No. 2 of Georgetown, were "the best team ever before a hose cart, either in Colorado or in any place else in the world," according to the *Miner*. The betting was heavy; the differences between the teams could be measured in fractions of a second, with the judges and their stopwatches often subjected to complaints. The "straightaway" race—the pure foot race in which contestants pulled a cart without dropping or connecting hose—was the hottest competition, with the Alpines as the favorites. The *Miner* copied the *Denver Republican*'s report, saying it was the only "unbiased" account the paper could find. Tensions ran high:

> Bets were at first even, and finally enthusiastic and ardent admirers, who believed the mettle was in the [Alpine] boys, offered and gave odds of 5 to 2 they would beat 27¾ or any time made on the course during the day. So when everything was ready excitement ran high. If a stroke of lightning had started the boys they could not have gotten off quicker or evener. Will Reilly and Billy Williams, on the lead, saw nothing or heard nothing after the word but the chalked line, to them then the dearest and greatest object to reach of their existence, and it was so with the whole team. They seemed to be fairly inspired to fly and to win that race, and they did. Never before in Denver, or at any firemans' tournament in the State were such shouts heard. From the second they started they seemed to grow into the entire multitude's heart, as their choice, and not a man, woman, or child upon the ground failed to yell, scream or shout till their throats were sore, as the team came rushing down the track. The first and fast friends of the crew yelled and flung their hats, caps, and even to taking off their coats in order to express and show their delight. All this excitement seemed to fill the boys with encouragement who exerted themselves more and more as they neared the line, gaining instead of losing time as they neared the end. With the air ringing with shouts the Alpine Hose Company, of Georgetown, crossed the chalked line with the time of 27¼ seconds, the best on record. Some of the boys were actually carried by friends around the track by the crowd. For ten minutes it was all a man could do to hear himself speak so great was the excitement. Chief

Duggan, who is a member of the Bates Hose company [Denver], rose up in his buggy and in a loud voice offered to bet any man

ONE HUNDRED THOUSAND DOLLARS

that the Bates team could beat the Alpine [Bates previously had a time of 27¾ in this race]. For a time every one stood silent at this great offer of a bet, the largest assertion from a reliable man ever heard on a public ground or race course. Mr. Hamill [of Georgetown] stepped up to the Chief after a time, and told him whenever he was ready to please step up to the Windsor Hotel and he would find money to back the Alpines for any amount for $1,000 and upwards.

No one took up the bet, and William Hamill came away the hero of the day for the young Alpine team. The frenzy of the crowd and the dollars overwhelmed the occasion, however, causing the *Miner* to note: "We rather think that we have seen the last of State Firemen's Tournaments."[27]

Georgetown's business district became a regional center. Two of the town's leading Jewish merchants had branch stores. Sam Strousse's dry goods business grew to include branch stores in Silver Plume and Leadville. Louis Cohen opened new stores in Silver Plume and Red Elephant. The town's Jewish community was small, well respected, and apparently well integrated with its gentile neighbors. The Strousse children were welcome guests at a birthday party for Lillie Hamill, daughter of the influential William A. Hamill.[28] The death of the Cohen's young daughter in 1882 was noted with great sorrow throughout the community.[29] The Jewish community included immigrants from Russia, Germany, and Poland.[30]

The town's Jewish merchants generally closed their stores in observance of Yom Kippur and Rosh Hashanah, sometimes traveling to Central City to join others in celebration or even going to Temple Emanuel in Denver.[31] The customs were generally received with a combination of curiosity and respect on the part of the community. Locals seemed intent on trying to understand some of the Jewish traditions, even sponsoring a lecture:

> Are you familiar with the peculiar customs of the Jewish people? Do you know how beef must be killed and prepared to be eaten by an Orthodox Jew? What are some of the differences between the Reformed and Orthodox Jews? Ever see a phylactery, and know what the wearing of it means? Come and hear the lecture at the Methodist Episcopal church next Thursday evening, where these and many other things regarding Jews will be spoken of and explained. The lecture will be illustrated by Jewish paraphernalia. The spirit of the lecture is most kindly, and anybody can enjoy Mr. Potashinsky's presentation of his subject. The lecture is free, and an offering will be taken. Don't miss "The Sons of Abraham, or the Modern Jew."[32]

The town's willingness to listen and learn seemed to extend to all but the Chinese and Indians. Georgetown's early residents knew about and supported Colonel John M. Chivington and his attack on the Indians camped at Sand Creek in 1864. The word *massacre* never entered the discussion, only the need to protect settlers from attack. One early meeting on Indian policy resulted in a public resolution in response to "the expressions of misplaced sympathy now being uttered in the east, through the press, towards the Indians of the West."[33]

By 1878 the rhetoric had taken a turn for the worse. The *Miner* reported on a band of Utes camped west of Empire. The headline read "UGLY UTES— Their Devilish Disposition Developing." The article claimed that Indians had attempted to take possession of Junction Ranch in Tabernash.[34]

In February 1879 Governor Pitkin called on men throughout the state to serve as needed, appointing William Hamill a brigadier general in the First Division of the state militia.[35] Hamill seemed nonplussed. As the *Miner* reported: "General William A. Hamill has received his commission as Brig. Gen. He is modest about it, however, and doesn't carry it around with him."[36] The local press, however, seemed pleased that someone would be watching over Middle Park to assure that nothing would sweep down on the town from the north.

Feelings ran hot when White River Utes killed Indian agent Nathan Meeker and ten associates on the Western Slope in September 1879. Hamill's actions were followed closely: "General Hamill sent over this morning 500 ball cartridges, and has telegraphed Governor Pitkin for 50 stands of arms with 40 rounds of ammunition. With these the men in Middle Park think they will be able to protect themselves if attacked."[37]

Most of the children in town cowered at the sight of Indians, generally small bands of Utes who occasionally traveled through the area. Longtime resident Sarah Rowe Corbett later wrote of her early memories of Indians in Georgetown:

> During the summer, when I was a little child, the Indians used to come to town; the squaws with their papooses on their backs. They would come to the open windows of our home and say "bisquit papoose" meaning they wanted food for their babies. My mother would always give them food, but my little brother and I would hide under the bed. Our idea of the Indians was that they would scalp us.[38]

The Chinese in town were not feared but were reviled. Given any opportunity the local press made sport of the Chinese population, referring to them as "celestials" or other epithets. The presence of opium dens did not sit well with the locals, nor did the fact that the Chinese did not attempt to westernize. The *Miner* printed many lengthy articles on Chinese immigration, most complaining of the number of immigrants and their unwillingness to blend into American culture:

> In the "E pluribus unum" of our country, however, is an immiscible element. This is the Chinese race. It is, and must for ever remain, an extraneous ingredient in our population, having no affinity for our customs and usages, and possessing no interest in the welfare and perpetuity of our government. . . . It is their immense numbers that render them formidable and dangerous. China has a population of five hundred millions, in round numbers, an

average of four hundred and sixteen inhabitants to the square mile. The United States contains less than fifty millions, exclusive of Indians and Chinese. It will thus be seen that the whole country could be swamped at any time by the unrestricted invasion of the Asiatic horde. The increasing influx of this undesirable people, which has been temporarily checked by the resistance of the working classes in California, is liable to assume the form of a human deluge within a few years, particularly on the Pacific slope, sapping the vitality of our industries and checking our substantial progress. Their aggression in other countries may justly be viewed as prophetic of what may be expected in our own.[39]

The editor's curiosity did lead him to the back room of a Chinese laundry to view their style of religious worship:

Adjoining the city hall [in the Alpine Hose building] is the "washee house" of Soe Long. In one end of the front room is a tabermach or altar beautifully and exquisitely ornamented with artificial flowers. Upon the altar is an idol before which jossticks are burnt. There is also a glass lamp suspended that is continually burning. Here then is the shrine at which the Georgetown Celestials worship, but they do not stop at this, but "alle samee likee Mellican man," they "fight the tiger," and within the sacred precincts of the altar. To try to describe the game is an impossibility, for the intricacies would drive to distraction the oldest sporting man in town.[40]

Despite the fears of being overwhelmed by Chinese immigrants, Georgetown's Asian community was small, limited to several families operating Chinese laundries. The 1885 Colorado State Census indicates that only eleven Chinese were living in the town.

By 1885 the town's largest group of immigrants was the Irish, with Swedish, English, and German groups close behind. Swedish immigrants had started to appear in large numbers in the late 1870s, establishing a Lutheran congregation that met in the Congregational church on July 8, 1877, with well-respected local resident Rev. Reuben Weiser officiating.[41]

Two weeks later, services were announced. The sermon would be in English, but both German and Swedish "altar services" were provided.[42] The Lutheran congregation bought the Congregational church in 1881, with services soon performed in Swedish by the first permanent minister, Edward Nelander.[43]

The variety of languages and accents continued to fill the streets of Georgetown through the 1880s. The Germans incorporated the German Benevolent Society, as well as the local Turnverein Society.[44] The *Miner* spoke highly of one of the group's late winter celebrations:

Among the most respected of Georgetown citizens are those who belong to the German element. They are always in the vanguard of enterprise, and are ever liberal when asked to contribute for the sake of charity. That they should have their socials and meetings in society together is nothing but natural, and that they always enjoy themselves is a fact not disputed. The MINER was honored with an invitation to the social party gotten up by the Germans of this city, on last Tuesday evening, the event being in celebration

KJÆRE SKANDINAVISKE BRØDRE !

I det Haab, at de Skandinaviske Brødre i Georgetown og Omegn med Glæde ville samles om Guds Ord, forkiyndt i Fædrenes Sprog og i Fædrenes Renhed, tillader Undertegnede sig at bekjendtgjøre, at der Lørdag Aften Klokken 7-30 og Søndag Eftermiddag Klokken 4 vil blive afholdt Gudstjeneste i den

Kongregationale Kirke

i Georgetown. Kjære Landsmænd og Brødre! I ere alle hjertelig velkomne!

O. ASPERHEIM,
Georgetown, 6 Juli, 1875.   Norsk-luthersk Præst fra Wisconsin.

Call for Lutheran congregation. Courtesy, Denver Public Library, Western History Collection, Randall Collection.

of the German holiday known as "Fasnacht." The scribe found one of the happiest gatherings he ever saw in Georgetown, and enjoyed the pleasantries of conversation and splendid singing of those present. The entertainment was supplemented by a supper that was in keeping with the other appointments of the evening and enjoyed by all. After supper dancing and singing again filled the programme of pleasure until the early hours of morning witnessed the farewells to one another of the joyous gathering.[45]

The town's Irish residents organized several groups, ranging from political action (such as the Fenian Brotherhood and the Irish Land League) to military (the well-known Emerald Rifles, complete with gray uniforms with green trim) to social groups (the Ancient Order of Hibernians). The highlight of the year was St. Patrick's Day. In 1886 a grand parade was held:

The procession of last Wednesday formed at the catholic church and headed by the Georgetown band playing Irish national airs, marched to the west end of Alpine street, where they met the delegation from Silver Plume, consisting of the Ancient Order of Hibernians in full regalia. Following the order were others wearing badges of green. The two processions were then consolidated and marched to the notes of "wearing of the green." First in line came the Ancient order, commanded by Mr. Wood, following were the scholars from the Sisters' school, and they by a large number of ladies and gentlemen residents of this city and Silver Plume. There were 250 in line and [they] presented an imposing picture as they wended their way to the church. At the church Solemn High Mass was celebrated by Rev. Father Carmody, assisted by Fathers Maloney and Turner. The ceremonies were very impressive and lasted about an hour. After mass, Rev. Father Maloney delivered an address, in which he pictured the past and present condition of Ireland, graphically reviewing her sufferings. His address was replete with eloquence and was a strong argument for the rights of the Emerald Isle, as well as an earnest appeal to all Irishmen to remain true to their native land.[46]

Even the smaller ethnic groups hosted events for all. The Welsh celebrated St. David's Day on March 1, and the Italians celebrated St. Barbara's Day in December.

Perhaps the two best indicators of general progress and civility were the establishment of schools and a hospital. Both public and private schools were started in the earliest days of the mining camp, with institutions serving both white and black students.[47] The highlight was the construction of a beautiful two-story brick school building in 1874. After initial concerns about the structural stability of the architect's design (a Mr. Lewis), the school opened to the delight and pride of the local community.[48]

In the fall of 1880 the Catholic church announced that it would be building a parsonage and opening a school in the basement of the church, to be administered by the Sisters of Loretto.[49] Whereas many of the state's churches worked to open parochial schools in the early years, most were located in metropolitan areas. The Sisters' School in Georgetown stood as a great success in a mountain community. By the fall of 1882 a new brick school and the long-awaited hospital were well under way, completing the Catholic church complex along Main Street.

The Catholic hospital opened its doors in 1880 in its temporary home, Demain House in lower town. In September 1883, after three years of fund-raising and construction, the new brick structure opened to patients. Local mining companies and individuals donated to the cause. Miners who paid fifty cents per month would be treated at the hospital at no additional cost.[50] The two-story building would provide care for all. The Sisters of St. Joseph managed the facility, which was gratefully received by the community:

> The Sisters have begun moving this week into their new building. It is really and [sic] admirable affair, with light rooms, high ceiling, hard walls, hot and cold water, registers from the cellar furnaces, gas and every convenience that modern architecture can suggest. The first floor will be used by the Sisters; the second floor devoted entirely to patients. Two halls, running east and west and north and south, respectively, divide each floor into four sections. Entering at the front door, which faces west, one finds a small reception-room and the main stair case, which is spiral, on his left. North of these, opening upon the north and south hall, are the Sisters' dining and sitting rooms, two in all. These constitute the northwest section. On the right of the main entrance are two large reception-rooms, connected by folding doors,

which occupy the southwest section of the first floor. The southeast section is taken up with a large sleeping room for the Sisters, and north of it a back stairs, with a stairway behind or east of them leading to the two furnaces in the cellar, only one of which, however, will be used except in extreme cold weather. In the northeast section are a large kitchen, with range, boiler, sink, pantry and dumb-waiter to the patients' rooms above; a large laundry with three stationary wash-tubs and a bath-room. The main hall leads into Griffith mountain on the east, and a commodious cellar for provisions and coal has been constructed there on the same level with the first floor.

In the second story there is a balcony over the front door, where the patients can sit and view the town. To the left of the hall, entering from this balcony, is a small room, corresponding to the small reception room below, which is to be used for the dispensary. North of it are a single and double room. The southwest section comprises three large front rooms. Opposite, in the southeast section, is a small room and a fine, large double one. The latter is to be used as a smoking and reading room for the patients. It has a commodious closet for a library and is very cheerful. The northeast section has a small room, a bathroom and a large double room, which is to be used as a ward. The dumb-waiter communicates with this. From the east end of the upper hall a plank causeway will be built to the mountain side so that patients brought on a stretcher can be taken in without the jar and pain of being carried up the stairs.

Every convenience is provided for effective nursing. All sick persons of every creed, no matter how poor, can be treated there. The rich can come and will be equally welcome. In fact several eastern invalids of means have applied for quarters, preferring them to a hotel. More Protestants, by a large percentage, have thus far been treated than Catholics. The Sisters know only the one word SICKNESS and that Charity bids them cure it. All else is left to the generosity of patrons and the public. Our citizens should give them earnest, most hearty and substantial support.[51]

Records of St. Joseph's Hospital show treatment rendered for everything from the common cold to silicosis, a deadly disease suffered by most hard-rock miners who worked underground for extended periods. Miners inhaled great quantities of rock dust suspended in the air after blasting or use of powerful drills. The sharp edges associated with hard-rock dust left the lungs torn and tattered, creating a deadly cough that gradually led to death.[52] The mortality rate grew steadily with the number of years worked in the mines.[53]

By the mid-1880s, unemployment and mine closures further complicated the lives of everyday miners. Although most of the larger mines continued to operate, many investors who had poured money into the local mines moved on to the larger operations around Leadville or other new strikes. The mineral existed in the mines around Georgetown and Silver Plume, but dollars for processing, acquisition of new equipment, and development of new technology became scarce.

Although the populations of the surrounding communities continued to grow, Georgetown started to stumble. New construction had ground to a halt by 1885. The *Silver Standard* [Silver Plume] described the problem in 1886:

Silver Plume and Idaho Springs are the only two towns in the county that are advancing any in the way of permanent improvements. Georgetown has seen only one new building in the last four years, and that is the hook and ladder house which is now almost completed. Up here new buildings are going up all around, and business is progressing more rapidly according to population than any other mountain town in the State.[54]

For the first time, the local press began to comment on the naysayers, pointing out the need to stay positive and encourage growth and development. Banker Charles Fish, a longtime resident, started to complain about the economy. In an effort to secure the money in his Bank of Clear Creek County, Fish refused to make loans on mining ventures, a move that infuriated local companies. The *Silver Standard* published some of his negative comments:

When asked if we were producing some ore he would say "Oh, they say we are; but it costs $2 to get $1 out." The sooner such croakers "cash in" and quit the game, the better it will be for Georgetown, but it is understood that the law allows them "three score years and ten," and they usually stay until "time" is called on them.[55]

The *Miner* further summed up the situation:

From 1872 to 1876 property in this town was a mint of wealth to anyone who was so fortunate as to own a building; the rents charged were exorbitant; every shed that proved a shelter against the elements [was] at a premium, and even old barns were transformed into dwelling houses. All people were prosperous; money was plentiful; poverty was unknown and every countenance beamed with happiness. From some cause unknown and incomprehensible all this underwent a change and a sense of depression took possession of us. Real estate took a tumble and suddenly dropped fifty per cent in value and continued in its downward tendency until it reached seventy-five per cent below its apprized value of 1886.

Everybody was at sea as to the cause. Our mines were yielding their usual abundance of wealth, and there was no visible cause for such a sudden change. Still, the fact remained that the spell was upon us. A feeling of insecurity pervaded the minds of all.[56]

As longtime boosters of Georgetown, the *Miner* editors had to end on a positive note. The article went on to say that good times were now at hand, that Georgetown would shine as never before—"Happy Days Are Here Again." Although there was "no visible cause" for the drop in property values, the lack of investors and the move to other, richer mineral districts did not bode well for the Silver Queen. In a letter to the editor, R. O. Old, an early supporter of the mining industry in the county, echoed the *Miner*'s sentiments:

Idaho Springs is going ahead! Silver Plume is going ahead! But is Georgetown moving forward? Are our neighbors, above and below us, following in a "rut"? I trow not. But Georgetown, our once lively Georgetown, is in one and follows it. I take it for granted that we all know, or should know, what is meant by the word "rut"—to wit: that, lacking public spiritedness, we are not abreast with the times and the world; that, having made no advance, or but little, we have become [an] old fogy, and like communities, old and young, that have passed away before us, we are dying of inanition and "dry-rot."[57]

The Georgetown Knights of Labor (Assembly 1, 3682) announced their first annual ball in December 1885. The establishment of a reading room in early 1886 came with the announcement that thirty new membership applications were being processed.[58] The downturn in national economics, combined with fear of continued mine closures and wage cutbacks, spurred local interest in labor issues and organizations.

Major investors disappeared from the scene. The price of silver continued to drop. The *Silver Standard*'s honesty stood in contrast to the boosterism of the Georgetown papers. The *Plume* editor painted a clear picture in 1886: "Nearly every mine owner in Clear Creek county at present [is] doing little more than keeping even with their property, hoping every day for silver to advance in price. . . . [S]ome of our best producing mines have been closed down, and will likely remain so until there is a favorable change in the price of silver."[59]

Indeed, the average price of silver dropped below a dollar an ounce in 1886, far from the peak of $1.40 in 1866. The value of the white ore had decreased steadily from the early days of the silver camp; nevertheless, production continued on a relatively steady pace, with actual production of silver in the county increasing in 1886. The residents of Georgetown wrestled with feelings of despair and discontent. Prices were down, but the industry seemed secure—why was the town struggling? Why were local mines closing?

Two factors came to the forefront: politics and the cost of transporting and treating the ore. Both of these factors directly affected investors' willingness to put their money into the mines. The fate of silver rose and fell with the federal government's decisions relating to coinage and the purchase of silver. From the panic regarding the demonetization of silver in 1873 (the "Crime of '73") to the celebrations related to the passage of the Bland-Allison Act in 1878 (requiring government purchase of silver), the townspeople acknowledged the relationship between the government and their livelihood.

More than any other community in Clear Creek County, Georgetown's fortunes and future fluctuated with the national economy. As noted earlier, Silver Plume continued to grow while Georgetown struggled, even though both towns depended on the same silver mines. Most of the mines continued to produce, but profits decreased. The miners in Silver Plume continued to work, extracting the ore as always; however, mine managers in Georgetown and their investors back East and overseas pulled back, taking their money to other endeavors. Silver Plume continued to house more and more miners (especially new immigrants coming west to join friends and families), whereas the monied interests in Georgetown started to move on. Some investors turned to the goldfields of Empire and Idaho Springs, expecting the drop in the value of one metal to result in a rise in the other or, perhaps, simply wanting to avoid the seeming vagaries of the silver market.

After the rise and fall of the Greenback Party in the 1870s and the passage of the Bland-Allison Act in 1878, the politics of silver quieted until the first call for a silver convention in 1885. D. B. Harris of Georgetown called for a meeting in Denver to discuss the need to remonetize silver. Over 1,000 delegates from several western states and territories attended.[60] From then on, all western mining interests began to focus on the industry's needs. An unusual coalition of mine owners, investors, businessmen, and miners—rich men and poor men, all with lives wrapped up in the continued production of silver—began to pull together to attempt to secure the future of the white metal. The future of Georgetown, as well as other silver communities throughout the western United States, lay in the balance.

NOTES

1. *Colorado Miner,* January 15, 1881, 3.

2. Ibid.

3. For information on the Bainbridge hanging, see the *Colorado Transcript* (weekly) [Golden City], May 1, 1867, 2; *Georgetown Courier,* August 3, 1907, 4.

4. For information on the Schramle hanging, see *Colorado Miner,* December 15, 1877, 3; Lafayette Hanchett, *The Old Sheriff and Other True Tales* (New York: Margent, 1937).

5. *Colorado Miner,* April 17, 1873, 4.

6. *Colorado Miner,* July 20, 1878, 3.

7. *Colorado Miner,* November 25, 1882, 2.

8. *Colorado Miner,* May 30, 1874, 3.

9. *Georgetown Courier,* November 8, 1883, 3; see also *Georgetown Courier,* September 13, 1883, 3.

10. *Georgetown Courier,* May 8, 1884, 3.

11. *Colorado Miner,* January 28, 1882, 2.

12. Ibid.; *Colorado Miner,* February 11, 1882, 3.

13. *Colorado Miner,* February 25, 1882, 3.

14. *Georgetown Courier,* April 6, 1882, 3.

15. Colorado State Census as transcribed by the Foothills Genealogical Society in its publication *Colorado Genealogical Chronicles* 15, Combined 1885 Clear Creek County Colorado State Census (Lakewood, Colo.: Foothills Genealogical Society, 1885).

16. *Colorado Miner,* January 28, 1882, 2.

17. *Colorado Miner,* May 20, 1882, 3.

18. *Colorado Miner,* April 8, 1882, 3.

19. *Colorado Miner,* February 25, 1882, 2. General Canrobet's quote was actually "C'est magnifique, mais ce n'est pas la guerre," according to a Crimean War Web site: www.geocities.com/Broadway/Alley/5443/cirmopen.htm.

20. *Georgetown Courier,* October 18, 1883, 3.

21. *Colorado Miner,* September 27, 1879, 2.

22. *Colorado Miner,* December 27, 1879, 3.

23. *Colorado Miner,* May 26, 1877, 3. For information on President Grant's visit, see *Colorado Miner,* October 16, 1875, 3.

24. *Colorado Miner,* November 17, 1883, 3.

25. *Colorado Miner,* October 16, 1875, 3.

26. *Georgetown Courier,* December 21, 1882, 3.

27. *Colorado Miner,* August 14, 1880, 2.

28. *Georgetown Courier,* May 12, 1881, 3.

29. *Colorado Miner,* July 15, 1882, 3.

30. Colorado State Census as transcribed by the Foothills Genealogical Society in *Colorado Genealogical Chronicles.*

31. References to Jewish holidays are found throughout the newspapers, since the merchants would remind local patrons that they would be closed. For further information on Jewish history in Georgetown and Colorado in general, see Ida Libert Uchill, *Pioneers, Peddlers, and Tsadikim* (Boulder: Quality Line, 1957); Allen D. Breck, *A Centennial Hisotry of the Jews of Colorado 1859–1959* (Denver: Hirschfeld, 1960).

32. *Georgetown Courier,* August 4, 1906, 4.

33. *Colorado Miner,* April 14, 1870, 4.

34. *Colorado Miner,* February 15, 1879, 2.

35. *Colorado Miner,* August 31, 1878, 3.

36. *Colorado Miner,* March 1, 1879, 2.

37. *Colorado Miner,* October 4, 1879, 2.

38. Sarah Rowe Corbett, "Reflections," unpublished ms., no date, 23. Available at Historic Georgetown, Inc., Georgetown, Colo.

39. *Colorado Miner,* December 18, 1880, 2.

40. *Georgetown Courier,* February 8, 1883, 3.

41. *Colorado Miner,* July 7, 1877, 3.

42. *Georgetown Courier,* July 21, 1877, 3.

43. *Colorado Miner,* February 12, 1881, 2.

44. *Colorado Miner,* February 19, 1881, 3.

45. *Colorado Miner,* March 13, 1886, 3.

46. *Colorado Miner,* March 20, 1886, 3.

47. *Georgetown Courier,* April 17, 1884, 3.

48. "All About a Foundation," *Colorado Miner,* August 15, 1874, 3.

49. *Colorado Miner,* September 4, 1880, 3.

50. *Colorado Miner,* August 14, 1880, 3.

51. *Colorado Miner,* September 1, 1883, 3.

52. Hospital ledger, St. Joseph's Hospital, Georgetown, Colorado. Photocopy available at Clear Creek County Archives, Georgetown, Colorado.

53. For more information on silicosis and troubles faced by hard-rock miners and their families, see Leanne Louise Sander, " 'The Men All Died of Miners' Disease'; Women and Families in the Industrial Mining Environment of Upper Clear Creek, Colorado, 1870–1900," Ph.D. diss., University of Colorado at Boulder, 1990.

54. *Silver Standard* [Silver Plume], October 9, 1886, 5.

55. *Silver Standard* [Silver Plume], December 14, 1889, 3.

56. *Colorado Miner,* June 12, 1886, 3.

57. *Georgetown Courier,* August 4, 1887, 3.

58. *Colorado Miner,* January 16, 1886, 2; February 13, 1886, 3.

59. *Silver Standard,* August 28, 1886, 4.

60. See *Georgetown Courier,* July 29, 1893, 3, for retrospective, as well as the *Colorado Miner,* March 27, 1866, 2.

Silver Plume's baseball team, champions of Clear Creek and Gilpin counties, 1889. Back row, left to right: Wm. N. Cliff, K. Beane, J. C. Connelly, Ed Overfield, Frank Gison, Will Guard, Dan Cavanaugh, John Nankivell, Harmon Noyes, Tom Cowell. Front row, left to right: Frank Pollard, Bruce Johns, Jim (Mickey) Coughlin. Courtesy, Verona Chappel Zimmerman.

Baseball

"[Baseball is [t]he American national sport, and no outdoor amusement or exercise equals it in genuine healthful recreation. As a means of invigorating manhood and developing the physical powers of youth, it stands without a rival" (*Rocky Mountain News* [Denver], April 16, 1867). It could do all that and boost civic pride as well, but baseball could also break hearts.

Georgetown took the game to its "bosom," as Victorians loved to say. A victory upheld town honor, and "fielding a nine" showed that Georgetown had become civilized. Yet pride and money rode on every game, and defeat hurt, as this excerpt from a local newspaper indicates.

> Sunday the Idaho club came up and proceed to put the boys in the soup to the tune of 14 to 8. The Georgetown club is composed of home boys and when they are defeated it's a home defeat. The

Idaho club is composed partly of professionals from nowhere in particular. The club boasts of having to pay three of their players $300 per month.
(*Georgetown Courier,* July 4, 1889)

Neighboring Silver Plume took defeat no better. When the local team was defeated by a score of 13 to 12, the *Silver Standard* (May 26, 1888) blamed guess who? "The umpiring was rotten," a missed a call in the eighth inning had allowed two runs to score. "The score should have been 12–11 in favor of the Plume." Still, sportsmanship won out: "there was but little kicking."

With teams playing wherever a level spot could be found, on an unmanicured diamond, and, until the mid-1880s, without gloves, scores sometimes ran as high as the hills. Fans cheered their team and, in the immortal words of Ernest Thayer, may have groaned when "mighty Casey has struck out."

Colorado Miner office. The person at the right is probably E.H.N. Patterson. Courtesy, Denver Public Library, Western History Department, X-1473.

—— Newspaper Wars ——

As important as the press was to its existence, a town the size of Georgetown, only briefly topping 3,000 residents in the 1870s and 1880s, hardly appeared able to sustain two newspapers. Yet throughout most of that period the *Colorado Miner* and *Georgetown Courier* fought for dominance, annihilating several smaller papers along the way.

In an era of outspoken, personalistic journalism, the two papers scratched and nipped like two kittens. Neither surrendered an inch until the *Miner* finally gave up the ghost in December 1887. A few examples from the *Miner* will suffice.

> Our esteemed contemporary accuses us of stupidity in saying that the comet is approaching the sun. Other stupid people, like the astronomer Richard A. Proctor, [say the same] . . .
>
> All the same we would rather be stupid with Proctor than as brilliant as our esteemed contemporary, who has in his lifetime been called a damphool even. (*Colorado Miner,* October 14, 1882)

> The popinjay editor of the *Courier* . . . (*Colorado Miner,* June 24, 1882)

The *Courier* [is] an alleged newspaper with about one-third the circulation of the MINER. (*Colorado Miner,* August 21, 1886)

Sitting on the sidelines, Silver Plume's *Silver Standard* (October 16, 1886) got in a few digs about the fight: "First round—blood for the *Miner.* Second round—*Courier* gets in a stab below the belt. Will fight to finish during the hot weather. Weapons—gust, slush, venom and spite."

The Albert Nelson house on Rose Street, built in 1881, complete with flowers in the yard and trees along the street. Courtesy, Denver Public Library, Western History Collection, X-1104.

Plants

Bedding plants at the green-house on Second street, include Alyssum, Antirrhium (snap-dragon), Asters, including New Rose, Schiller, Washington and Victoria varieties, Balsam, Centaurea, Canna, Calendula, Catch-fly, Clarkia, Dianthus, Datura, Eschsoltzia single and double, Godetia, Gaillardia, Gilia capitata and tricolor, Glaucium, Lobelia, Lychnis, Myosotis (forget-me-not), Nasturtium, Nolana, Poppy, Perilla, Pansy, Phlox, Petunia, Scabiosa (mourning bride), Stock, Verbena, Ainnia. Prices, with few exceptions, range from 25 to 75 cents per dozen.

—*Georgetown Courier,* May 14, 1885

THOUSANDS OF PEOPLE LIVED IN GEORGETOWN between 1859 and 1896. Although the recorded population never exceeded 3,500, many residents came and went as opportunities arose in other towns. Some families sent pictures to relatives who carefully noted names on the originals. Others sat for portraits now unlabeled and unknown. Some of the town's personality can be found in these portraits—they depict people of all colors, ages, situations, and circumstances.

The glimpses are by no means complete—many well-known families and individuals are not included here—but we have attempted to open the door and look back at some of the faces in the crowd.

Faces in the Crowd

Proud mothers could not resist a trip to the photographer's studio to show off their offspring. Above: Dorton Forbes. Right: Robert Montgomery Wheeler. Courtesy, Historic Georgetown, Inc., Wheeler Collection.

Again, proud parents dressed their young ones in appropriate finery for photos to send to relatives in the East. Above: James and Hazel Burkholder, ca. 1892. Right: William Hamill III and Harry Hamill, ca. 1890. Courtesy, Historic Georgetown, Inc., Hotel de Paris and Hamill collections.

Surveyor George Marsh, left, *and George Aiken,* right, *pose in front of Marsh's office near the corner of Biddle and Ninth streets. The building in the upper left corner of the photo is the Selak Brewery. Courtesy, Historic Georgetown, Inc., gift of Edith Dick.*

Above: *Hiram A. (Alex) Spruance served as chief of the Georgetown Volunteer Fire Department between 1878 and 1879. Courtesy, Harriet Commors Bellmyer and Marjorie Connors Lawrence, copied to Clear Creek County Archives. Left: The chief's uniform was more ornate by the time Chief Reuben B. Weiser (local dentist) posed for his portrait. Weiser served as chief between 1880–1881 and 1882–1883. Courtesy, Denver Public Library, Western History Collection, Randall Collection, F-4593.*

Above: *From the back of the picture: "Jake Snetzer—Tailor & very much beloved."* Right: *The Stanton family (managers of Barton House in the early years) had to squeeze together to fit into the photographer's studio. Courtesy, Denver Public Library, Western History Collection, Randall Collection, unnumbered.*

Above: *Portrait of five Chinese men taken in Georgetown. As identified by Jesse Randall,* "*Wa Chin* (center) *wears a long shirt and wide-brimmed hat; Tang Ya-Shun wears a suit and watch chain with a charm.*" Below: *Chas. Gow is the Chinese man in the cart with the pit bull at his side.* From left to right: *Jo. Raymond, unknown man, Fred Aiken, Jo. Cuerto, Paddy Welch, Geo. Layden, Chas. Gow, M. Welch, and two unknown men. Courtesy, Denver Public Library, Western History Collection, Randall Collection, X-21660, X-6549.*

The town's young women donned their best dresses for the family album. Upper left: *Gertrude Nash Pulsifer.* Upper right: *Probably Lena Strousse.* Lower left: *Lou DeLamar. Courtesy, Denver Public Library, Western History Collection, Randall Collection, F-4607, F-4619, F-5200.*

Clara Brown, generally known as "Aunt Clara," invested in mining property around Georgetown along with other African Americans in the late 1860s and early 1870s. Her niece, Josephine Smith, married local barber Charles Townsend. Courtesy, Colorado Historical Society, F-3714.

Few photographs of the town's early African American residents have survived through the years; most of those that do exist are unidentified. Above: *Albert McKinney photo of African American man, taken at "Georgetown Photograph Gallery at the Georgetown Vanity Store, Rose Street, Georgetown, Colo."* Left: *Unidentified African American baby. Courtesy, Denver Public Library, Western History Collection, McKinney photos, donated by Mrs. Humfreville, Idaho Springs, X-21532, X-21534.*

Above: *Joseph A. Love, builder of the Courier building, Love's Store (northwest corner of Seventh and Taos streets), and other buildings around Georgetown, became known as the town historian in later years. Courtesy, Colorado Historical Society, 10031804.* Right: *Louis DuPuy started with a small restaurant in Delmonico's Bakery, which he developed into the elegant Hotel de Paris, his "souvenir of France" in the Rocky Mountains. Courtesy, Colorado Historical Society, Dalgleish photo, 10031805.*

Above: *Dr. Irving Pollok (often misspelled "Pollock") served in the Second Colorado Regiment prior to becoming a local physician.* Courtesy, Colorado Historical Society, F-50,292. Right: *John Coulter, local attorney.* Courtesy, Historic Georgetown, Inc., donated by George S. Richardson.

Joseph W. Watson, local mine manager and entrepreneur, whose connections with Philadelphia investors and belief in the value of local mines helped Georgetown grow and prosper. Courtesy, Colorado Historical Society, F15,630.

Above: *Young women of Georgetown ca. 1890. Top row,* left to right: *Dot Price, Fannie Owen, Annie Hancock, Mollie or Bessie Perrin; center row,* left to right: *Addie McCreedy, Lillie Hamill; front row,* left to right: *Hattie Reynolds, Ella McCreedy, Esther Hancock, Mattie Bruce, Susie Fisher.* Left: *Schoolteachers from the Georgetown School, 1884. Left side,* top to bottom: *Hattie McCrimmon, Ida Parker, Bella Wright; right side,* top to bottom: *Addie McCreedy, Jessie McArthur, Ella McCreedy. Courtesy, John Tomay Memorial Library, Georgetown.*

Mr. and Mrs. Frank A. Maxwell posed in their home on Fourth Street, a rare interior view. Courtesy, Frost Family Collection.

8 No Longer the Queen

THE YEAR 1878—WHAT A YEAR IT WAS! COLORADANS WOULD SELDOM SEE ITS LIKE again. Not since those fabled days of the 1859 Pike's Peak gold rush had they seen so much anticipation, excitement, and mining fever. Almost overnight, it seemed, although in reality it did not happen quite that fast, Colorado changed from a gold to a silver mining state. People talked of millions, not mere thousands, in production, and millionaires strode the streets of Denver and Leadville. By 1880 Colorado would become the number-one mining state in the United States. The dreams of the fifty-niners had become reality.

All this came about because of Leadville, tucked away at 10,200 feet in the upper reaches of the Arkansas River. Leadville, only two passes and several mountains away from Georgetown, took away the latter's title of Silver Queen in a twinkling. In 1879 Leadville mines produced $11 million in silver; the amount was $14 million the next year. Leadville did not have to worry about railroad connections. Railroads raced to tap the profits awaiting them. Almost before it had reached full stride, Leadville gained that key to mining prosperity, which Georgetown had yearned for and finally achieved well past its youth.

Even the Comstock, now fading from the Big Bonanza days, gave way. Nor could some of the new districts such as Montana's Butte, Arizona's

Tombstone, Dakota's Deadwood, and Colorado's San Juans even start to dim Leadville's star. Leadville was "queen" without par.

Leadville's excitement and wealth attracted newspapermen and journalists. Their reports did much to make the town spectacularly famous almost overnight. Writer and artist Mary Hallock Foote went there with her mining husband, Arthur. "All roads lead to Leadville. Everybody was going there!" she wrote, and that they did in 1878–1879. What she saw amazed her: "Leadville was one of the wildest and also one of the most sophisticated of the mining camps." The instant city and its mines served as a mecca for many mining engineers, as Foote knew—as did Rossiter Raymond, David Brunton, and others who had reported on or been at Georgetown. Of Leadville Brunton observed: "At that time Leadville was the Mecca for adventurers of all kinds, and everything was crude. Leadville continued to grow with mushroom-like rapidity."

Leadville quickly became Colorado's second-largest community and even dreamed of taking the capitol away from Denver. Millionaires worked Fryer Hill properties with names like the Robert E. Lee, Morning Star, and Maid of Erin. They then took their wealth to Denver, jump-starting that community, which quickly became the Rockies' predominant town. Politically, those Leadvillites dominated the scene for the next decade and convinced Coloradans that their destiny rode with silver. Horace Tabor and his Leadville-enriched neighbors ambled confidently across the Colorado landscape.

Investors flocked to Colorado, having forgotten the gold bubble and other misadventures. Leadville both hurt and helped other Colorado districts. It initially took attention away from the San Juan mines, but that district finally came into its own, as did the Gunnison country and other smaller districts pulled along by Leadville. Prospectors dashed into canyons and climbed mountains, looking for a "second" Leadville. Eventually, they found one at Aspen.

Everything appeared larger than life and sired legends that made it seem even grander and more adventuresome than the hard reality and uncertainties of mining and living in isolated camps amid the grandeur of the Rockies. Once again, Colorado was the place in which to invest, to visit (maybe to sample some of the "sin" in these high-riding towns and camps), and to read about in newspapers and magazines.

What could Georgetown do about all this commotion and promotion? Not much. It might gain a little through the general interest in the area, some investors might wander by and an article or two appear, but nothing on the scale of Leadville occurred. Georgetown had once grabbed center stage by pushing rivals aside, but now it found itself in the wings. A new mining rush played no favorites; those who could not keep up dropped out of the race for public attention and investment.

Georgetown residents were not sheltered from such developments. The *Georgetown Courier,* for instance, carried an article from the *Lake County*

Reveille that discussed Leadville's "simply enormous ore production." According to the reporter, that came about only with "development" work.

Suddenly, the $2 million Georgetown's mines had produced in a single year seemed less thrilling. Further, Georgetown lacked the flush of youth, the excitement of the rich, new mines, and the allure to beckon investors to mines that appeared pale in comparison to the Chrysolite, the Little Pittsburg, or the Matchless. Certainly, some people would come, and Georgetown would bask in the backwater of the silver fever and Colorado's newfound fame. Nevertheless, the community and its mines were "old," and its mineral potential was better known and tested. Leadville produced far more in its first four years (1878–1881) than Georgetown could muster in the entire two-plus decades from the 1870s into the early 1890s.[1]

The impact on Georgetown had become noticeable by the spring of 1879, as Leadville and its near neighbor in Summit County, the Ten Mile District, sprinted into full swing. The *Courier* on July 3 summarized its "chief industry" after the first six months of the year "had glided by," concluding in its Victorian style:

> The greatest change that has occurred, which is noticed by the older residents, is the loss of many of our old miners who have been drawn away temporarily, to the newer camps, and consequently the development of good prospects has been checked to a considerable degree, and we shall probably feel the loss for some months to come before the matter is fully remedied.

"Temporarily" extended longer than locals and the editor had hoped. In the *Courier*'s January 1, 1880, summation of the previous year, the editor confessed: "The past year has not been noted for mining excitements in this county, the important new discoveries being few; and there is no question but [that] prospecting has been checked to a great extent by the exodus of a large number of our old miners to the new fields of Lake and Summit counties."

The census returns showed dramatically what had happened. In the 1880s Georgetown lost more than 1,200 people, nearly 37 percent of its population. Most of that decrease came in the first half of the decade.

The *Courier* put the best face possible on such developments. "Valuable properties" had been transferred to parties from other states. These people "of means" had purchased improved machinery and planned to, or were, working those properties on an "extensive scale." Fortunately, the paper stressed, a number "of our noted mines" paid attention to development work throughout 1879. The purpose for doing so was to keep a "fair amount of reserves" ahead of the "stopers," miners doing the actual work of digging out the valuable ore. Rather than "swell" production for the time being, such development work denoted "a very healthy condition for our mining interests," whereas "[y]ounger districts" tended to glory in large production statistics.

Further, Clear Creek County mines "are nearly all controlled by mining men, which is the principal reason why there are so few failures here." No business in the world "needs practical, experienced business men more than mining does." These last comments came in September 1880 after Leadville and the Ten Mile District had gone through a series of misfortunes and mine failures as a result of poor management, stock speculations, and overexpectation from a public that thought only wildly anticipated dividends would come from their investments.[2]

At Georgetown, optimism remained high. Coverage in the local newspaper seemed more focused and detailed than ever. Articles on individual mines and tunnels carried the location and the owner's name or names, along with an article a paragraph or more in length, not simply a brief notation.

The Richardson Tunnel on Leavenworth Mountain, for example, had lessees sinking a shaft to intersect the tunnel, along with doing, "at present," work entirely developmental in nature. The article discussed 1,500-an-ounce silver per ton. The Polar Star on Democrat Mountain was "progressing steadily," with the lessees raising "large amounts of quartz carrying some ore."[3]

The common theme among most properties throughout the period 1878–1880 was one of leasing. Even the noted Pelican-Dives property had lessees working along with the company's sixty to eighty men. Further, little 1,500-an-ounce ore reached the mills. That must have been a pocket, not a vein, but it did match stories coming out of Leadville. The report may have circulated with malice aforethought on the part of the company, the lessees, or the newspaper. More commonly, reports of around $40-an-ounce silver seemed promising. Whatever the value, miners were receiving less for their ore than they had a decade earlier.

The problem faced by Georgetown's silver miners, and those throughout the West, was the declining price of silver. This emotional "silver issue" will be discussed in Chapter 10, but it never languished far from Georgetown's collective conscience. The silver issue pitted East against West, debtor against creditor, and silverite against gold bug, while generating increasingly strong feelings and controversy.

The *Courier,* in its January 1, 1880, edition, noted that the "depreciation of the silver price" meant a direct loss to Colorado miners in 1879 "of about 1¼ million" dollars. At the same time, Coloradans aided "Pennsylvania iron men and eastern manufacturers" through the protective tariff with probably about the "same amount," which did not appear fair: "It seems to us that this whipping the silver miner at both ends is getting to be a little monotonous." The debate was just warming up in the press and in discussions.

The other problem, as the 1870s drew to a close, was the slowing down of prospecting and development of lesser-known properties. Comments such as the observation that Democrat Mountain "remains somewhat dull, but no worse than last year" balanced the report that a number of "new silver lodes" had lately been discovered on Lincoln Mountain or that Republican Mountain had more "activity prevailing this year than last."[4]

Reports in the *Engineering and Mining Journal* seemed even more optimistic than some of those appearing in the local press. The year 1878 provided some very positive statements for faraway readers. Consulting mining engineer Francis L. Vinton wrote a long article on the district and had much to say about Georgetown and the industry overall.

> It is without doubt the busiest mining area of the State, and Georgetown, the capital city, was never more active than at present. The mines on every side are in good pay, cheering the noble industry and stimulating the enterprise that demands so much intelligence, energy and fortitude. Whether highly skilled or not, the plucky miner is a man of quality to admire for nothing taxes the physical and moral strength more than this invasion of the depths where the resistances of Nature, though so great, are often less obstructive than her bewildering mysteries.

Vinton—a West Point graduate, Union Army veteran, and the first professor of mining engineering at Columbia University—continued his analysis of Georgetown mining: "It is true that great companies create great mines, and true, also, that certain mineral situations are best evolved on extensive plans, but such combinations, however convenient, are indispensable in a productive place like this."

After praising the fact that "more method is being practiced in mining," he enthusiastically cheered the "greater respect paid to experience, what we call mining science." Had this been the case earlier, some litigation and "much useless expenditure of money [would have been] avoided." Vinton was right on all counts. He understood what had happened too often in the district and elsewhere.

Vinton also emphasized that the western miner lived a better life than his eastern counterpart, the laborer. Furthermore, miners had come to accept that millions might not be their reward: "The genuine workers have come to know that they are better off than the average worker of the East and settled down to win their square human living from the veins of the earth, quite disabused of the fatuous idea that the name of 'Mine' is a charm to conjure millions."[5]

Vinton was correct in his view that the "genuine" miner, not the speculator, should work a property. That Georgetown needed companies with financial resources had been shown by the previous decade's results, although some locals still resented the idea that they could not do it on their own. Vinton was less enthusiastic about Georgetown's reduction capabilities, describing them as more of an ore market. This remained a problem of production and finances. Most local ores left by train for Denver's smelters.

Georgetown miners and others could not have been displeased with this honest, forthright evaluation of where they had been and the current situation. Vinton, however, had written his assessment just before Leadville exploded on the scene and where he himself would soon travel.

Vinton's observations must have intrigued eastern investors. The future seemed more promising than it had in years, and his article reversed some of the negative publicity that resulted from the Pelican-Dives mess.

The 1880s found Georgetown mining reaching its angle of repose, even if it was not what its boosters had hoped for. Georgetown had come to rest, and with it a generation's dreams of fame and fortune. The *Courier* faithfully supported its district and made the best of what it had. "Rich Strikes: Clear Creek County to the Front" trumpeted the headline of a July 1881 article. This was the "most important" strike in seven years, the article went on to say.

The *Courier*'s concluding summary of 1881 described the year as "prosperous," with work "carried forward without excitement, without the bluster of new districts." That little dig at Leadville and its contemporaries ended with this observation: a score of "valuable strikes" had been made in Georgetown, and if they had been in "newer districts" they would have "created a furor of excitement." Georgetown, however, took them in stride.

The *Colorado Miner* tolled the same theme. The editor, although mumbling about the end of the year "when it is our somewhat onerous duty to collect figures from our millmen and ore buyers," did so and thereby boosted Georgetown's image: "Amid the noise and bluster of surrounding camps she was almost lost sight of. The wheels of industry never flagged for a moment, however, but silently and surely Clear Creek County was pouring her metallic wealth over the land."

The *Engineering and Mining Journal* saw trends it praised, echoing Vinton and the *Miner*. Development work was going on in "a quiet way," and the people in Georgetown "have settled down to hard, honest, and determined work." Additionally, "those engaged in it, although not making much noise, nor indeed, any wonderful 'strikes,' are on the whole, doing very well." The reporter expressed true Victorian sentiments. Despite having heard of "wonderful good fortune" in distant mining camps, workers in Georgetown "believe that if they diligently woo the blind goddess here, she will not refuse to smile on them."[6]

During the 1880s, Georgetown basked in the afterglow of Leadville's and Colorado's mining successes. Leasing continued to gain a steady hold on local operations, although not always with good results for the property's owner. The Polar Star on Democrat Mountain, for instance, had produced early, then been leased. A report in the *Engineering and Mining Journal* stated that it had been "badly worked by lessees and will require considerable expenditure of money to put it in paying shape." That was being done, the article concluded. The Pelican-Dives, on the other hand, continued to be operated by both the company and lessees, with apparently fair results.

The *Courier* persisted, hammering on the new districts with their "public excitement" and "extravagant prices." Investors were advised to avoid them in favor of the "districts temporarily neglected," where good mines could be "secured for low prices" and "reasonable expectations [could be] realized."[7]

Outside investors proved a two-edged sword. Georgetown and Colorado needed them, but matters did not always turn out as expected or hoped. The

Courier, in October 1881, praised the eastern parties who developed carefully, with no publicity, and who "now are making money." Alas, not all mines did so, however: "The mines that are now staggering under loads of debts were all paying properties while under the management of Colorado men. The new owners had to learn from experience, which they have paid for." The ever encouraging editor thought the new owners had learned and "apparently profited by" it.[8]

This love-hate relationship was typical in the mining West, whether in a new or an old district. It carried over to railroads, the federal government, lawyers—the list seemed almost endless. Nothing migrated west without strings attached, and nothing was ever free, despite seeming appearances.

One thing Georgetown could do to counteract the new districts was to herald its own new discoveries and strikes. In its well-named "Drift and Dump" column, the *Colorado Miner* described such discoveries as the Broad Gauge, a yet-unnamed lode near Grizzly Pass, and the "new find" near the Magenta Mine. A new strike in the Joe Reynolds or any of its counterparts also received coverage. The "biggest strike of the present season" or the "latest discovery" might never be heard of again. Nevertheless, they grabbed the spotlight for a moment and helped promote their district.[9]

Reports on Georgetown became routine in the *Engineering and Mining Journal* as the 1880s progressed. Brief notes about this mine or that tunnel, mostly copied from the local press, were all that the district rated, however. No special reports or reporters examined the district. That left promotion and news dissemination up to the *Miner* and the *Courier,* and neither paper shirked its duty.

A report in the *Courier* (December 27, 1883) covered a variety of topics in what once had been the annual year-end summary. The feature appeared sporadically now, in part because mine owners and companies were reluctant to give the newspaper the information it needed. In a rather humorous vein, the reporter reasoned:

> We have stolen a number of reports on properties that were not "newspaper mines," and others that "it was none of our d— business" to know about, but could not steal enough to make a fair report. An annual report of prospect holes [which freely report] would not show well for Clear Creek County, which contains some of the most valuable productive mining properties in the state.

The article went mountain by mountain, selecting several mines and tunnels to discuss. For example, the paper reported excitedly that the Sonora Mine on Griffith Mountain had started a new tunnel. The lessee had undertaken to run a tunnel to open the lode, therefore "there certainly must be something there to warrant the undertaking. We hope he will strike it big." On Sherman Mountain work had resumed on the Illinois Mine, and on Republican Mountain the Wisconsin Mine was improving as miners reopened an old caved-in section of the workings.

Fascinatingly, in 1883 tunnels had again caught Georgetown's attention. Lessees, who had hit a four- to five-inch pay streak, were working the old Lebanon Tunnel. The Dives Company opened a new tunnel, and the Diamond Tunnel had reached 2,188 feet in length. The usual stories also appeared. Two men were working the Wolverine Lode, and an association of businessmen from Newport, Kentucky, appropriately planned to work the Blue Grass Lode. The next year the mining news shrank to less than a column, with no coverage at all some weeks. The political excitement of the 1884 state and national campaigns pushed local mining off the pages, which said more than was likely intended about Clear Creek mining.

Georgetown did not ignore a wonderful promotional opportunity in the early 1880s: Denver's National Mining and Industrial Exposition, bankrolled by Horace Tabor, among others. Old-timer Robert Old, "one of our most public spirited citizens," took charge of the Clear Creek exhibit. The display of local ores "surpasses any in the building," the *Courier* proudly decided.[10]

Inclement weather and bad roads continued to hamper local mining. So did melting snow when the water eventually trickled downward, flooding some mines and causing "the pumps to be kept busy" and expenses to mount.

Through it all, Georgetown and Clear Creek County continued to look to the future. Voltaire, with a touch of cynicism, once noted that optimism "is a mania for saying things are well when one is in hell." Years ago, "those were mushroom times, destined to quickly wither under the suns of a few months." In 1885 Clear Creek "has arrived at a season of substantial prosperity. It is not of a spasmodic, but [a] solid nature." All this had occurred "quietly without excitement and hurrah."[11] Such optimism was epidemic in western mining regions.

The *Courier*'s editor also envisioned eastern "moneyed men" arriving this spring because, he reasoned, "no locality in the state offers better inducements." "Quite a number of important mining sales this summer" would surely follow, he believed. Unfortunately, Georgetownites lived on hopes. What they commonly read were comments such as one asserting that the Terrible Mine was assuming its former "good condition."

Some basis for hope persisted. Production in Georgetown remained in the high $1 million to low $2 million range throughout the decade. The one difference in county production reflected the fact that until the start of the 1880s the Georgetown District had been the principal producer. Now the Idaho Springs and Freeland districts gradually increased their gold production and their share, aided to a smaller extent by Dumont and Lawson. Even some small placer operations reminded old-timers of twenty or so years before, when Colorado placer mines ruled the territory.[12]

Part of the new gold emphasis came about because of the declining price of silver. This created more interest in prospecting for gold and encouraged development of marginal mines that contained both metals.

2d SILVER BENEFIT.

McClellan Opera House,

One Night Only.

Saturday Evening, April 3d.

THE ONLY AND ORIGINAL

BUFFALO BILL

HON. W. F. CODY,

Late Chief of the Scouts of the U. S. Army,

and his MAMMOTH COMBINATION
in his great Sensational Drama, entitled,

"The Prairie Waif,"

Introducing the Western Scout and Daring Rider,

Buck Taylor, King of the Cowboys.

A Genuine Band of Pawnee Indians,

Under Pawnee Billy, Boy Chief and Interpreter.

24 First Class Artists.   **New and Beautiful Scenery**

Mr. Cody, "Buffalo Bill," will give an exhibition of fancy Rifle Shooting,
holding his rifle in twenty different positions, in which he is acknowledged preeminent.

Prices of admission as usual.   Reserved seats, one dollar,
to be had at Forbes & Stromberg's.

Silver benefit handbill, 1886. Courtesy, Denver Public Library, Western History Collection, NS-244.

One hope, the *Miner* correctly saw, rested with low-grade ore. As George Hearst's toast to his famous Homestake Mine in the Black Hills proclaimed, "Here's to low-grade ore and plenty of it." A district always had more of the less glamorous low-grade ore, but what a difference it could make if it could be profitably milled. Although he believed Clear Creek County would still yield plenty of high-grade ore, the editor of the *Colorado Miner* looked to the future: "What her future output from the leads and lodes of low grade will soon be, imagination even can hardly compute."[13]

Although newspapers did not like to report about the dangers of mining, which were bad for the district's image, brief comments did appear. Accidents happened even in the best-regulated and most well-developed mines, although—fortunately—no major disasters had occurred in Georgetown. Nevertheless, danger was part of the profession. The *Miner,* discussing the death of a local miner, reflected the vicissitudes of the profession. The miner, the paper eulogized, had joined a long list of miners killed, "which reminds us of the perils of the miner's life. So many noble fellows have been taken from among us."

Sadly, lessees often took fewer precautions as they worked to develop mines and make profits within a limited amount of time in properties that were not theirs. Falling rocks, premature explosions, bad air, cave-ins, falls down shafts—the list stretched on as men labored in the dimly lit recesses of the earth. "Miners' consumption," or silicosis, caused by rock dust filling the lungs, also took a steady toll. Those steam and electric drills beloved by owners and stockholders were the prime culprits.

Miners who were single simply tramped on to another district as their predecessors and contemporaries had been doing for generations. Married men had more responsibilities and were less footloose. None, however, could avoid the hazards of their profession. Once, there had been the hope of making their own strike and becoming owners. Now, with corporate control gaining ascendancy and new discoveries fewer and farther between, an era was ending before their eyes.

As the decade came to a close, the hopes and dreams persisted and, just as persistently, ran up against reality. Reports mounted: the Pelican-Dives property "recently was sold under a mortgage sale," the sheriff sold the property of the Astor Alliance Mines, and the Centennial Mine was "sold under a trustee's sale." The Pelican-Dives finally reincorporated and continued on as a producer. The newspaper speculated that a tramway from Georgetown to its mines might be as successful as the Gilpin County tramway. Local members of the Mining and Business Men's Association might look into that, the *Courier* suggested.[14]

As it turned twenty-five, Georgetown had already lasted longer than many of its contemporaries. The district had clearly seen better days, but so had much younger Leadville and neighboring Central City. Clear Creek mines still produced at a steady pace—creating jobs, generating sales for local businesses, and serving as the basis for the town's existence. The years 1859 and 1864 had long faded into history, and the twentieth century was now much closer than those bygone days.

Thus as the 1880s ended, Georgetown had not regained its former prominence but had managed to maintain itself as a mature district with continued production. People had tramped out of Georgetown to pursue their dreams in newer, more exciting districts. In the words of the old "Tramp Miner's Song":

Up this road I've been before
No one will ever know
And I missed the path
And I can't go back
And no one will ever know.

NOTES

1. M. H. Foote to Helena [Gilder], May 12, 1879, Hague Collection, Henry E. Huntington Library, in Rodman Paul, ed., *A Victorian Gentlewoman in the Far West* (San Marino, Calif.: Huntington Library, 1972), 200; Brunton interview in T. A. Rickard, *Interviews With Mining Engineers* (San Francisco: Mining and Scientific Press, 1922), 73–74; *Georgetown Courier*, December 26, 1878, 1; Charles W. Henderson, *Mining in Colorado* (Washington, D.C.: Government Printing Office, 1926), 109, 176; Summary Colorado Census Returns, author's files.

2. *Georgetown Courier*, July 3, 1879, 1; January 1, 1880, 1; September 4, 1880, 3.

3. *Georgetown Courier*, May 30, 1878, 3; June 6, 1878, 3.

4. *Georgetown Courier*, July 3, 1879, 1; quotes from January 1, 1880, 3; March 26, 1880, 2.

5. Clark Spence, *Mining Engineers and the American West* (New Haven: Yale University Press, 1970), 32, 38–39, 339; "Georgetown," *Engineering and Mining Journal*, April 20, 1878, 271.

6. *Georgetown Courier*, July 28, 1881, 1; December 29, 1881, 1; *Colorado Miner*, January 1, 1881, 1; *Engineering and Mining Journal*, October 15, 1881, 256.

7. *Engineering and Mining Journal*, July 9, 1881 27; quote from September 30, 1881, 174; September 11, 1882, 257; *Georgetown Courier*, April 20, 1882, 3.

8. *Georgetown Courier*, no date, quoted in *Engineering and Mining Journal*, October 15, 1881, 256.

9. *Colorado Miner*, July 9, 1881, 3; August 6, 1881, 3; June 23, 1882, 1; July 22, 1882, 1.

10. *Colorado Miner*, March 17, 1883, 2; October 13, 1883, 3; *Georgetown Courier*, January 4, 1883, 1; quote from July 26, 1883, 3; December 27, 1883, 2; January 3, 1884, 1; *Engineering and Mining Journal*, April 22, 1882, 212; May 13, 1882, 250; May 23, 1883, 305; November 10, 1883, 299.

11. *Georgetown Courier*, January 8, 1885, 1. See also January 10, 1884, 3; May 22, 1884, 1.

12. *Georgetown Courier*, April 23, 1885, 2; Henderson, *Mining in Colorado*, 32, 109.

13. *Colorado Miner*, October 13, 1883, 3.

14. *Colorado Miner*, November 10, 1888, 2; *Engineering and Mining Journal*, February 25, 1888, 149; July 7, 1888, 10; August 11, 1888, 113; quotes from November 3, 1888, 377; January 5, 1889, 17; July 6, 1889, 12; August 3, 1889, 100; *Georgetown Courier*, September 13, 1888, 3.

Dalgleish photo of Southgate's home, "Hermitage," complete with merry-go-round. Courtesy, Historic Georgetown, Inc.

Chimney Rock?

The rock formation behind the county courthouse became known as Chimney Rock after Joseph Watson and associates built a mill with a large brick smokestack perched atop the formation. In the 1890s one of the town's best-known characters, Benjamin Franklin Southgate, lived at the foot of the rock, where he had a merry-go-round and a cabin with this sign: "Independent Scientific Repair Shop, Clocks Coopering Chairs Saws Filed, Tin Ware Politics and Theology Tinkerd."

In 1893 Southgate asked the town board to change the name Chimney Rock to Bunker Hill. Board member Ernest Le Neve Foster, a Briton, pointed out that many foreigners were living in the town, some of whom did not think of Bunker Hill in a positive light. The board discussed the matter:

> Selectman Parker thought that Foster ought to be excused from voting on the question, and [Selectman] Harris believed that foreigners should not be allowed to vote anyway. After indulging in

their joking sallies, the motion of Selectman Harris was carried, and the name "Chimney Rock" is to be relegated to oblivion and Georgetown's center piece, and Uncle Ben Southgate's monument foundation is to be known as "Bunker Hill."

(Georgetown Courier, February 6, 1892)

Carrie Everson

Dr. William Everson was one of those souls who invested in Colorado mines with little mining knowledge. The predictable happened. In the early 1880s he and his wife, Carrie, visited Georgetown to see what had happened to their money. While there, the problem of economical reduction of low-grade silver ore fascinated Carrie Everson.

Born in Illinois in 1844, she married Everson, a Chicago physician, twenty years later. Growing interested in his work, she studied "various branches of science" and became "proficient in chemistry." Following her trip to Colorado, she studied mineralogy. After testing various ideas, Carrie developed a revolutionary process that became known as flotation, which she patented in August 1885. She described what she had found as "the chemical affinity of oils and fatty substances for mineral particles."

The couple tried in vain to find support for the process in Georgetown or anywhere else. Unfortunately, her husband died in 1889, and Carried never received the backing she needed to try her idea. She eventually returned to Denver and took up nursing. Her loss was the industry's loss as well.

INCORPORATED TOWN, COUNTY SEAT, RAILROAD destination, tourist attraction, business center— all these terms described Georgetown by the time the railroad came to town in 1877. The completion of the Georgetown Loop to Silver Plume in 1884 furthered the town's reputation. Over the next decade, even under the clouds of economic uncertainty surrounding the value of silver and mining in general, Colorado's "Silver Queen" continued to mature. The number of new buildings decreased, but those that were built stood as outstanding examples of Victorian architecture.

The Wild West was in the town's past, barely a part of its memory.

Jewels of the Silver Queen

GEORGE TOWN, COLO. HIGH BRIDGE IN DISTANCE, C. & S. R.R.

Railroad track winds around Georgetown with the loop bridge in the distance in this photo ca. 1886. The Methodist congregation decided not to chance having sparks from the locomotives strike their building, and they moved the structure to Taos Street in 1884. That option was not available to the Catholic church, with its large brick structure—something the parishioners might have regretted when the building burned in 1917. Courtesy, Lee Behrens Collection, A. Rapin photograph.

Undated photograph (ca. 1890) of the interior of Grace Episcopal Church, showing gaslights, wood-burning stove, and profuse floral displays—possibly for Easter Sunday. Courtesy, Grace Episcopal Church Foundation, Burkholder Collection, George Dalgleish photo.

George Dalgleish photo of the Presbyterian church on Taos Street, Georgetown (also known as "the old stone church"). Courtesy, Denver Public Library, Western History Collection, F-1419.

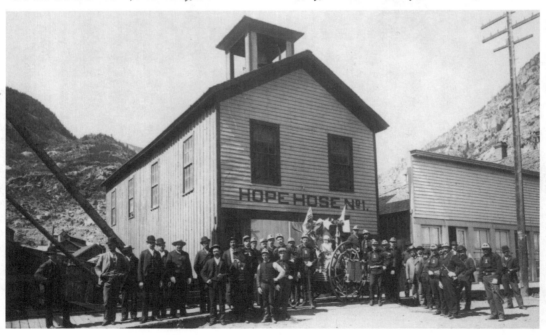

The first fire house built by the town of Georgetown, later known as Hope Hose No. 1, was built on the south side of Sixth Street spanning Clear Creek. Courtesy, Colorado Historical Society, George Dalgleish photo, F-15672.

Priest, nun, and children in front of Our Lady of Lourdes Catholic Church and rectory on Main Street. Photo is labeled "Father Howlett's Church and house." Courtesy, Denver Public Library, Western History Collection, F-644.

Above: *The William A. Hamill residence, ca. 1885, with new additions to the main building, granite wall, fountain, office building, and carriage house—a fine home for the "gentleman from Clear Creek" and his family. Gift of Roland Pilz to Historic Georgetown, Inc.; reprinted courtesy, Denver Public Library, Western History Collection, X-1259. Below: Georgetown's brick schoolhouse, the design of architect "Mr. Lewis" and built by James Cree in 1874, stood as the pride of Georgetown in this 1890s photograph. The caption reads "Horton Lyon 2nd from left." Courtesy, Historic Georgetown, Inc., Alex Martin photograph, gift of James R. Fitzpatrick.*

Alpine Hose No. 2, the second hose company building with its distinctive fire tower added to the original structure in 1880. Courtesy, Colorado Historical Society, George Dalgleish photo, F-15671.

Above: *The Star Hook and Ladder Company's building, now the Georgetown Town Hall, built by David Duncan in 1886 at a cost of $1,750, is shown decked out for the Fourth of July. Courtesy, Denver Public Library, Western History Collection, Conway photo, X-1378. Below: Clear Creek County Courthouse, ca. 1890 (originally the Ohio Bakery), at the southwest corner of Fifth and Argentine streets, with the brick addition added in 1887. Courtesy, Colorado Historical Society, J. B. Stanton photo, CHS-X-7217.*

Father (later Bishop) Nicholas Chrysostom Matz supervised the construction of both of these fine buildings for Georgetown's Catholic congregation in 1882. The Sisters' School" (above) stood on the west side of Main Street across from the hospital (below). Courtesy, Denver Public Library, Western History Collection, X-1203 (photo above); Lois Keck photo, X-1094 (photo below).

The town waterworks, shown in this photo ca. 1885, provided piped water—a valuable commodity for the convenience of town residents and the safety of their homes and businesses. Courtesy, Colorado Historical Society, Bivans Collection, Mellen photo, F-40498.

Cushman's three-story brick building towers over the rest of Georgetown in this view to the southwest from Biddle Street between Eighth and Ninth, ca. 1895. Courtesy, Denver Public Library, Western History Collection, Strohmeyer and Wymon photo, X-1339.

The home of Frank Maxwell, known for its ornate details and elegant tower, received its final facelift in 1891 under the ownership of Mr. B. F. Potter. The Georgetown Courier (November 7, 1891, 3) noted that the changes "will make it one of the handsomest [houses] in the town"—and it still is. Courtesy, Frost Family Collection.

By the early 1890s Georgetown boasted new brick buildings in the commercial district and several new residences in the latest styles. Only the vagaries of the national economy loomed as clouds on the horizon. Courtesy, Historic Georgetown, Inc.

The brief economic revival around 1890 gave Georgetown some of its best-known architectural landmarks. Above: John Henry Bowman completed his new residence at Ninth and Rose streets in 1892. Courtesy, Historic Georgetown, Inc. Below: Louis Dupuy finished the Hotel de Paris in 1889. Courtesy, Colorado Historical Society, 10026521.

9 A New Coat of Paint and the Last Hurrah:
The Silver Queen
Enters the 1890s

HOW TO MIX PAINTS FOR TINTS.
Mixing red and black makes brown; lake and white, rose; white and brown, chestnut; white, blue and lake, purple; blue and lead color, pearl; white and carmine, pink; indigo and lampblack, silver gray; white and lampblack, lead color; black and venetian red, chocolate; white and green, bright green; light green and black, dark green; white and green, pea green; white and emerald green, brilliant green; purple and white, French white; red and yellow, orange; white and yellow, straw color; white, blue and black, pearly gray; white, lake and vermillion, flesh color; umber, white and venetian red, drab; white, yellow and venetian red, cream; red, blue and black, olive; yellow, white and a little venetian red, buff.
—*Georgetown Courier,* August 1, 1891, 3

THE PASSAGE OF THE SHERMAN SILVER PURCHASE ACT IN 1890 ADDED NEW LIFE TO Georgetown, similar to the changes the new "tints" would make to the old houses formerly painted white with green trim, as referenced by early visitors. Between 1890 and 1892 the quiet streets of Georgetown, stilled by closures of local mines and the lack of new construction, sprang back to life. After eight years of little growth, the town suddenly experienced a building boom. The next generation of businessmen stepped up to build their legacies.

Even Charles Fish, the stodgy banker previously heard bad-mouthing Georgetown on the streets of Denver, built a handsome two-story brick building, the new home of the Bank of Clear Creek County. The old wooden building at the corner of Sixth and Rose had burned in April 1887. Fish immediately announced plans to rebuild, then lagged along until times took a turn for the better. The Georgetown correspondent to the *Silver Standard* [Silver Plume] commented on the lack of progress and the need for a progressive attitude: "The old hunkers that now cumber our town must either give way to live business men or retrograde will be the order of things until that is accomplished."[1]

Georgetown's main commercial district—the blocks along Sixth Street—underwent major changes during this time. Henry Kneisel, Henry Seifried,

Elizabeth Curtis, and H. A. (Alex) Spruance tore down the old Spruance and Hutchinson Store in 1891, along with the old Curtis Hardware Store the next year, to build two new brick buildings on the south side of Sixth. Charles Fish built another brick building adjacent to the bank property, which was soon leased to the Georgetown Masonic lodges.

Elegant residential properties appeared at the same time. Thomas Cornish's distinguished brick home on Argentine (1892) and John H. Bowman's Italianate house on Rose Street (1892) reflected their owners' belief in the future of Georgetown and the area's mines, as did B. F. Potter's major renovations to the old Baily and Nott home on Fourth Street (later known as Maxwell House) in 1891. At the same time, Louis Dupuy expanded the Hotel de Paris again, with another new wing on the east side of his property—changing the exterior to brushed concrete and topping off the design with a statue of Lady Justice (without the customary blindfold) holding the scales of justice. The fact that Dupuy commenced his improvements (1889) prior to the passage of the Sherman Silver Purchase Act (1890) indicated his confidence in the town's growing tourism business.

Within two to three years the town's appearance changed. New, well-designed buildings replaced older clapboard structures. Homes were improved and painted in the newest colors. The town added flagstone sidewalks to the downtown and acquired property for a city park. Things were well under way by 1890:

> The Park commissioners are busy selecting designs for landscaping the grounds. It is really interesting to note the taste and skill of some of those who have submitted drawings to the commissioners. Mr. H. H. Atkins and Mr. A. R. Forbes have each submitted a design that is creditable to the taste of a professional landscape gardener. Others have likewise made submission of suggestions which we have not yet inspected. It will not be easy to choose from so much really excellent material but the plan selected will doubtless be that one that meets the most general approval.[2]

The town fathers took the enterprise seriously, carefully studying the designs, then spending their own time constructing the new town project. Mayor Kneisel, Albert Forbes, and members of the town's Board of Selectmen volunteered hundreds of hours to the project.[3] At the same time, the town undertook repairs to the sidewalks and planted young cottonwood trees:

> The town looks twice as lively as it did only two years ago. There are now hundreds of handsome, flourishing shade trees, well built walks, new buildings and a general extension of improvements that promises great things for the coming year, and all because a few leading citizens and city fathers determined that we must have shaded streets and finally a park.[4]

It was not just a park, but a well-designed area with carefully planted trees, walks, and a fountain. The *Courier* challenged: "What other mountain town

Centennial Celebration

——OF THE——

INAUGURATION OF GEORGE WASHINGTON

THE FIRST PRESIDENT,

——AT THE——

Presbyterian Church, Georgetown,

APRIL 30th, 1889.

PROGRAMME.

Reading of Proclamations of the President, Governor
and Police Judge,
REV. W. E. KNIGHT.

Anthem by Choir.

Services of Episcopal Church used 100 years ago at the
Inauguration of Washington,
REV. EDGAR.

Address, "Washington, the True Democrat,
JUDGE L. H. SHEPARD.

Music, Instrumental.

Address, "Religious Character of Washington,"
REV. E. F. FOWLER.

Singing, "America," by the congregation.

Address, "Common School Education, the Bulwark of
our Liberties,"
C. C. POST, ESQ.

Address, "Our Country,"
JACOB FILLIUS, ESQ.

Reading Selections from Washington's Inaugural,
A. E. CHASE, ESQ.

Music, Quartette.

Reading Selections from Washington's Farewell Address,
REV. FATHER LEY.

Benediction, REV. A. W. DAHLSTEIN.

*Centennial celebration of the inauguration of George Washington and class of 1890
program. Courtesy, Historic Georgetown, Inc.*

in Colorado has a park enclosed with iron fence and surrounded with stone flagging? Hurrah for Georgetown!"[5]

The last major town improvement came in 1891 when the Electric Light Company (of Georgetown) was granted a twenty-year franchise for streetlights—spurred on by the construction of a new hydroelectric plant on South Clear Creek—and consistent delivery of electrical power to the houses and other buildings in Georgetown. The company secured the water power rights from the Farwell mill site, just below Devil's Gate Bridge, then commenced construction of the plant next door to George Hall's lumber mill at the west end of Fourth Street.[6] As it requested a franchise license from the town, the company promised to deliver power to Georgetown and the surrounding mines, which would certainly be seen as an economic boost.[7] Shortly after receiving approval from Georgetown, the company made preliminary arrangements to deliver power to Silver Plume.[8] Construction of lines to Silver Plume would wait until 1893 and would require further negotiations.[9]

The Electric Light Company had first supplied limited power to Georgetown homes and businesses in 1886, when M. T. Morrell arranged to use the water power of the Clear Creek Mill at the east end of Fourth Street, on South Clear Creek.[10] The plant operated intermittently for the next few years, continually upgrading its equipment and adding new customers. The town dealt the company a heavy blow in 1887 when the contract for streetlights stayed with the gas company, even though the electric light proposal was considerably less expensive.[11] The political clout of the gas company, owned by banker Charles Fish, among others, outweighed that of the newcomers in the electric company.

The townspeople resented paying the additional cost for gas, and by 1891 many people stated a strong preference for electric lighting, causing the town to issue the twenty-year franchise agreement mentioned earlier. A permanent solution came in 1893 when the officers of the gas and electric companies joined forces to incorporate the United Power and Light Company of Clear Creek County.[12] Electricity rapidly became the preferred power source for home, business, and mining use, causing the gas plant to cease operation on September 1, 1894.[13]

As the town settled in to life with modern conveniences, the female residents started to focus on a statewide ballot issue to grant women the right to vote. The last time the issue had grabbed headlines in Georgetown was during the 1877 election when the town, as well as the newly formed state, resoundingly defeated the issue. Several of the women active in the 1877 campaign still lived in Georgetown and stepped forward to encourage passage of the issue. Their job was made easier by Mary Collins's success as a member of the local school board. Collins (Mrs. Dr. Collins in the press), the first woman to hold office in Georgetown, was elected to the school board in 1877 and had served with distinction over the years.[14]

Both Georgetown and Silver Plume formed Equal Suffrage associations, even though the *Courier* was convinced that women's suffrage was not of interest to Georgetown's female residents.[15] The Silver Plume chapter arranged for a lecture by Carrie Lane Chapman, much to the delight of the audience and the Silver Plume press:

> The exercises of the evening opened with a piano solo by Mrs. Smith, followed by prayer by the chaplain of the Equal Suffrage League, after which a chorus was rendered by a number of voices. Mrs. Kendall, president of the E. S. League, made a short address on the important question which confronts the people of this state, and then introduced the speaker of the evening, who made an impressive and convincing argument in favor of extending the right of suffrage to her sex. Mrs. Chapman possesses a fine voice and pleasing address, and she held the attention of her hearers from beginning to end with a flow of eloquence seldom equaled by the "stearner sex" and her points were roundly applauded. Mrs. Chapman no doubt made many converts to the cause among both sexes.[16]

The time had come—suffrage passed in November 1893, and one more change settled in on Georgetown.[17] Electric lights shone from the houses, arc lighting covered the street corners, and houses sported "the latest" colors. One change was not permitted, however: "A woman attempted the divided skirt fashion of riding horseback in Georgetown a few days ago, but was given just one minute to dismount. It's an innovation that our Marshall is not ready to accept until the custom becomes general."[18]

Times also changed for the mining industry. The public began to question the mining companies' use of public lands. Unbridled cutting of trees for lumber and fuel by mine owners began to create barren hillsides, subject to flooding and erosion. In 1887 the state of Colorado established a State Forestry Commission and sent a notice to all newspapers: "The growing scarcity of timber, the increased frequency of floods and droughts, the lack of water for irrigation, and other ills attending upon the destruction of forests, have become known to the people of this State, and an earnest endeavor is being made to remedy the evil."[19]

The last thing the mining industry felt it needed was any kind of government "remedy." Things remained quiet, however, until 1893 when the *Miner* announced the proposed establishment of forest reserves:

> If a forest reserve is to be set apart covering this county, it will withdraw from the public domain all the mineral lands in the county; no lodes can be located, and none can be patented. The mines which are now working will have to stop, as no timber will be available for mining purposes, and consequently our only industry will have to cease, and we shall have to allow the grass to grow in our streets until such time as congress passes a law, permitting mining within the forest reserves. When will this be? It may be years, and certainly not very soon. Our citizens had therefore better awaken to the fact at once, and protest.[20]

With headlines requesting that "Citizens Bestir Yourselves," the editor painted a grim future. Forest reserves were created in 1893, with the Pike Forest Reserve encompassing lands in western Clear Creek County, but this new program would be the least of the county's problems.

Fear of unemployment and problems prying paychecks out of local mining operations spurred the growth of local unions. Chapters of the Western Federation of Miners were strong and radical compared to the quieter ways of the earlier Knights of Labor. By 1890 the *Silver Standard* saw trouble on the horizon: "The annual troubles between capital and labor are now very active and strikes are on the tapis. The only solution of this question seems to be for every man to do his own work, and every man [to] work for himself."[21] In April 1892 Georgetown miners called a meeting to establish a miners' union.[22] Although discussions remained civil, the stage was set for a fight. Mine owners and local editors began to worry about what would happen to their struggling industry and recalcitrant investors.

NOTES

1. *Silver Standard*, March 17, 1888, 2.
2. *Georgetown Courier*, May 1, 1890, 3.
3. *Georgetown Courier*, May 29, 1890, 3.
4. Ibid.
5. *Georgetown Courier*, July 25, 1891, 3.
6. *Georgetown Courier*, April 18, 1891, 3; March 7, 1891, 3.
7. *Georgetown Courier*, February 7, 1891, 3.
8. *Silver Standard* [Silver Plume], April 18, 1891, 3.
9. *Georgetown Courier*, March 11, 1893, 3.
10. *Colorado Miner*, July 31, 1886, 3.
11. *Georgetown Courier*, August 11, 1887, 3.
12. *Georgetown Courier*, April 14, 1893, 3.
13. *Colorado Miner*, August 18, 1894, 3.
14. Diane L. Pope, "Friends of Progress: The Woman Suffrage Movement in Georgetown, 1877," unpublished ms., n.d. Copy available at the Clear Creek County Archives, Georgetown, Colorado. In 1877 the only precinct in the county to vote in favor of granting women the right to vote was the town of Empire.
15. Quoted in the *Silver Standard* [Silver Plume], October 14, 1893, 3.
16. *Silver Standard* [Silver Plume], October 28, 1893, 3.
17. Suffrage passed in all three Georgetown precincts: South, 104 to 53; West, 66 to 29; and East, 105 to 40. Silver Plume's two precincts split on the issue: West, 52 to 48; and East (losing), 51 to 78. Nevertheless, the issue passed in the county and, where it mattered most, at the statewide level.
18. *Georgetown Courier*, August 8, 1891, 3.
19. *Colorado Miner*, May 21, 1887, 2.
20. *Colorado Miner*, January 28, 1893, 3.
21. *Silver Standard* [Silver Plume], April 19, 1890, 2.
22. *Georgetown Courier*, April 9, 1892, 3.

Trees

The Council at a special meeting last Friday night discussed the questions of helping the citizens to improve the town by purchasing trees in large quantities and disposing of them at cost. A contract was let [to] James Buchanan to furnish 300 trees. They will be disposed of to any one who desires to beautify their yards or streets at 40 cents each. A hundred or more were spoken for within an hour after the Council decided to let the contract, and no doubt the trees will find ready purchasers as fast as they are delivered. If Georgetown had undertaken this little bit of improvement ten or a dozen years ago, the town would now be blooming like a garden. But it's never too late to commence, and three hundred trees started this season will give our people more courage to celebrate Arbor Day in 1890 by planting a thousand.

—*Georgetown Courier*, April 18, 1889

Young cottonwood trees line Taos and Eighth streets surrounding the Georgetown school. Wooden boards protect the young trees from wandering cows or other animals apt to chew on the bark or snap the young saplings. Courtesy, Denver Public Library, Western History Collection, X-1127.

Georgetown Volunteer Fire Department racing team, ca. 1896. Front row (left to right): 1. Will Schaur, 2. Lou Campbell, 3. George Old. Second row (right to left): 4. Charlie Campbell, 5. Bert Jarrell, 6. Ed Nelson, 7. Tom Pollard, 8. Roy Blackman, 9. Charlie Seibert. Third row (left to right): 10. Henry Blatter, 11. Harry Hendershot, 12. Bob Weaver, 13. Joe Lindsey. Back row (right to left): 14. Emil Anderson, 15. Ad Hinsdale, 16. Louis Parker. Courtesy, Georgetown Volunteer Fire Department, donated by Mrs. Josie Clark.

Urban Rivalries

As mining communities sprang up throughout the Clear Creek valley, urban jealousies and rivalries soon took root. Georgetown found itself at odds with Silver Plume, Empire, Idaho Springs, neighboring Central City and Black Hawk, and even Denver. In this "grow or die" world, much was at stake, and few punches were pulled.

Rivalries sometimes surfaced in snide comments, like one that appeared in the *Rocky Mountain News* [Denver] (November 6, 1868): "Speaking of Georgetown. I think I shall not go there, it is too much trouble." Or, commenting on the boosterism of 1878 with state pride, "It [Clear Creek County] is, without doubt, the busiest mining area of the state, and Georgetown, the capital city, was never more active than at present."

Civic superiority could surface in sports, as was seen when the baseball team won or Georgetown's Alpine Hose was given "Three Cheers and a Tiger" for defeating Denver, Boulder, and Central teams in a hose race. Comments were also made about the prettiest girls, the

fattest babies, or the most progressive businesses—
anything that would give the town an edge and get readers
thinking about how wonderful, prosperous, and grand
Georgetown was.

Newspapers carried the banner—promoting,
defending, attacking, and generally serving as a booster
chamber of commerce. A particularly aggressive, quotable
editor was worth his or her weight in gold.

THE COLORADO CENTRAL RAILROAD ARRIVED IN Georgetown on August 1, 1877, amid great fanfare and celebration. Tourism and mining interests smiled knowing they had iron rails (albeit narrow gauge) to connect them to the markets and people they needed to succeed. Extension of the line to Silver Plume in 1884 and to Graymont later that same year furthered these interests.

After the arrival of the train, talk quickly turned to the need to lower rates (especially for shipping ore) and extend services. A contentious relationship ensued among the line's various corporate owners and subcontractors: Colorado Central; Colorado and Southern; the Georgetown, Breckenridge, and Leadville; and the Union Pacific.

Arguments aside, the train provided the town with some of the most spectacular elements of its history. The Devil's Gate viaduct—or high bridge—fascinated railroad men, engineers, children, tourists, miners, and all who came to see it.

The Railroad and Devil's Gate

The terms far famed, serpentine, and engineering marvel appeared time and again in descriptions of the narrow-gauge route between Georgetown and Silver Plume. This Rose and Hopkins photo shows the track headed up the valley, crossing the creek, crossing over itself on the high bridge, then winding its way to Silver Plume. Courtesy, Denver Public Library, Western History Colleciton, H145.

The railroad companies and local promoters quickly commissioned photographers to develop brochures and pamphlets promoting the Georgetown Loop, such as this page entitled "A Day in the Rockies." Courtesy, Colorado Historical Society, Henry Buckwalter photo, CHS-B84.

William Henry Jackson shot some of the best views of the railroad and the valley between Georgetown and Silver Plume. This photo is shot from the curve known as the High Fill looking west to Silver Plume, with the Lebanon Mine and Mill complex in the valley to the right. Courtesy, Denver Public Library, Western History Collection, WHJ-415.

The view from the valley, looking back toward Georgetown, with the Lebanon Mine and Mill complex in the center of the photo. Courtesy, Lee Behrens Collection.

Above: *The Devil's Gate viaduct, or high bridge, spanning Clear Creek with the high bridge, Devil's Gate footbridge, and dam shown lower right, 1891. Courtesy, Historic Georgetown, Inc., W. Kilburn photo.* Below: *The loop valley looking north from the Hall Mill with the high bridge in the background. Courtesy, Historic Georgetown, Inc.*

Left: *Visitors by the hundreds posed for pictures on the high bridge—a daring feat on a windy day. Courtesy, Colorado Historical Society, CHS-B86.* Right: *The railroad companies and local promoters staged multiple trains in the same photo in an effort to impress potential visitors (and to have a little fun themselves). Courtesy, Harold Frost Collection, Alexander Martin photo.*

... HIGH LINE . GEORGETOWN TO SILVER PLUM

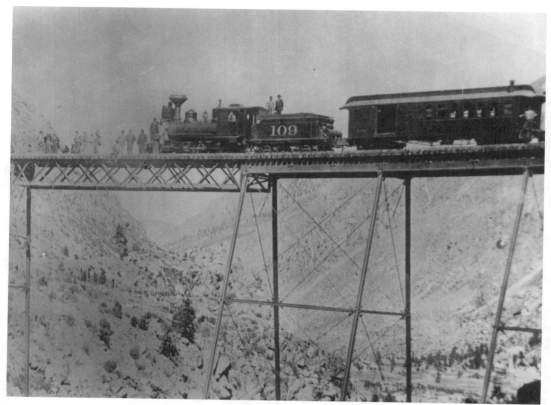

The true railroad buff can name the train, manufacturer, and other pertinent details (from donor's notes): 1895 on high bridge of Georgetown Loop. Cooke Feb 1884, #1550, Cyl. 15 × 18; Driv. 40"; Weight 58,300 lbs; To C & S #4; Former D S P & P #39. Courtesy, Colorado Historical Society, donated by P. A. Ronfer, 10031803.

Facing page, top: The tracks running alongside the Catholic church in Georgetown caused no problems for over thirty years but eventually sparked the fire that destroyed the church in 1917. Courtesy, Denver Public Library, Western History Collection, X-1421. Bottom: Georgetown's strong winds have been known far and wide, especially when they puffed hard enough to blow an engine off the track—"Derailed in a Hurricane, Feb. 4th, 1885." Courtesy, Harold Frost Collection.

Georgetown's brick depot was built through local fund-raising in time for the train's arrival on August 1, 1877. Local merchants and businessmen raised approximately $5,000 (including at least $700 from local mining magnate William A. Hamill) for purchase of the grounds and construction of the building. Two big celebrations were noted in the local papers: the arrival on August 1, 1877, and the sendoff of local boys shipping off for World War I. This photo may have been taken during the first celebration. Courtesy, Denver Public Library, Western History Collection, Chamberlain photo, X-1469.

A rare view, titled "D. W. Gamble and his men at Work putting in Dam at Devils Gate, Georgetown, Colorado." Courtesy, Harold Frost Collection.

Devil's Gate, ca. 1870, from further north, showing the general setting. Courtesy, Denver Public Library, Western History Collection, Chamberlain photo, X-19304.

The Devil's Gate bridge, local landmark and picnic destination, south of town prior to the construction of the dam and, later, the railroad to Silver Plume and the high bridge. Courtesy, Harold Frost Collection, Chamberlain photo.

The high bridge, dam, and footbridge represented the railroad, mining, technology, and tourism—major elements in the nineteenth-century history of Georgetown, Colorado. Courtesy, Denver Public Library, Western History Collection, McClure photo, MCC-3254.

10 "Until Time Calls a Halt"

THEY MIGHT HAVE CALLED THEM "GAY" ONCE, BUT THE 1890s OFFERED FEW HAPPY, carefree times for Georgetown. A mining town nearly thirty years old, with rare exceptions, found its best years behind, not ahead, of it. The days of being Colorado's Silver Queen had receded into memory, a short reign ended by Leadville a dozen years ago. Only a finite number of ounces of silver and gold lay hidden in the surrounding mountains, and the ore that was easy to recover had been mined out. Now, as the miners dug deeper, it cost more to recover less silver. Newer mining districts offered more glittering allure for investors, who moved on with the excitement.

With few exceptions (the Homestake Mine at Lead, South Dakota, comes to mind), mines and mining districts experienced a limited life. New, exciting discoveries lured investors, miners, and others away. Even glamorous Leadville, once Georgetown's rival, slipped as silver Aspen rose and golden Cripple Creek exploded on the Colorado scene. Almost single-handedly, Cripple Creek turned Colorado into the number-one gold-producing state in the United States. By 1897 the district started a run of from $10 million to $18 million in annual gold production—a run that lasted two decades.

Indeed, Georgetown and its mines had already lasted longer as a major mining producer than had many of its contemporaries. Local mines still produced, maintaining an average of $2 million annually until 1897. In 1890 Georgetown set an all-time record of $1.9 million in local production of silver and, for the county, a total of $2.9 million in production. Increasingly, though, a higher percentage of production came from gold and lead. Silver production reached all-time highs in 1893–1894—over 2 million ounces per year—as the district struggled to maintain itself. Therein lay the problem: the price of silver had collapsed. When Georgetown's mines opened back in the 1860s, silver hovered in the $1.30 range per ounce, dropping to 60 cents in 1893.[1] Thus the "silver issue" was born.

Nothing stirred up more emotions for Coloradans of this generation than the silver question, the silver issue, free silver, the battle of the standards, or whatever they chose to call it. Actually, the reason for the decline in silver's price was simple—overproduction against a declining market. With the Comstock's and Leadville's 1870s bonanzas, along with those in lesser districts such as Georgetown and Caribou, silver glutted the market. Only so much could be used as plate or jewelry, industrial uses remained small, and silver coinage throughout the world declined in use as country after country dropped silver to turn to gold and the gold standard. The reason for this change was simple. Gold carried a fixed price of $20.67 an ounce, whereas silver floated in value; and silver's lower worth made it less flexible for coinage. As countries switched to gold, the market for silver dropped significantly just as production spiraled to all-time highs.

In spite of these logical explanations, Coloradans saw plots, international conspiracies, villains, and a "crime" (discussed later in the chapter) as the cause of their problems. They were not alone. Other western silver-producing districts, suffering equally, saw the same things.

The reality that mining cost more as the mines sank deeper into the mountains and the grade of ore generally diminished, quickly cutting profits, played little in their thinking. Further, neither the railroads nor the smelters had proved to be the anticipated saviors in cutting costs, which only made folks angrier. The world and a host of scoundrels seemed to be against the honest, hardworking western miner to deprive him of his just rewards. His cause had been found.

The issue was both old and new. The framers of the U.S. Constitution realized the need for a sound monetary system. The 1770s and 1780s had produced anything but such a system for the new nation, as states had enthusiastically issued paper money that promptly plummeted in value. To alleviate the problem, the Constitution gave Congress the sole right to coin money and placed the country on a bimetal currency system based on gold and silver.

The problem became one of finding a stable market value for the two metals. Either one or the other was undervalued or overvalued in comparison to the other. Finally, in 1834 Congress set the ratio at sixteen ounces of silver to one ounce of gold. Unfortunately, that ratio simultaneously undervalued silver and overvalued gold, driving silver coins almost completely out of circulation for a time as people melted them for their silver and then sold it. Congress solved that problem in the 1850s by reducing the silver content of minor coins.

Then came the Civil War. Responding to wartime financial emergencies, the U.S. Treasury issued paper money, or "greenbacks." The result was inflation and paper money fluctuating in value in relation to gold. At the war's end, the business community wanted to return to the gold standard, or, as business leaders called it, "sound money." To them, paper currency represented "cheap money" because of the instability caused by its sometimes violent fluctuations in value below gold. Debtors, on the other hand, liked to borrow sound money and pay the loans back with paper, thereby gaining an advantage in the exchange. In reality, some wanted inflation, whereas the idea horrified others.

Meanwhile, the mint ratio of 16:1 did not reflect the market price, since silver was worth more on the open market. Western silver miners, therefore, did not sell bullion to the Treasury, and once again silver coins disappeared from circulation. Congress responded with what to its members seemed a logical plan and stopped coining the silver dollar—the country's main silver coin—in 1873. All hell would eventually break loose over the "crime of '73," but at the moment no agitation surfaced. The government now purchased only a little amount of silver for the smaller coins. For all practical purposes, the United States had joined the gold standard.

Miners and western politicians initially took no notice because of the world price. Alarmingly, however, the international price soon started to spiral downward because of the factors mentioned earlier. Only then did this appear to be a "crime," and the banner was raised. It would seem that a relatively few silver producers would not have much power, but they did. From Colorado's entry in 1876 through 1890, seven western states joined the Union, giving them political representation out of proportion to their numbers and population.

As the silver price went down, the silver spokesmen wanted two things— government purchase of silver and a guaranteed price based on the 16:1 ratio that placed silver at $1.25 per ounce. For the next twenty-five years that fight dominated western thinking.

The *Colorado Miner* clearly defined the issues as it saw them. "From the very beginning gold and silver" had been recognized as money, according to the newspaper. Since silver had been diminished, prices had fallen, business had stagnated, debtors had been oppressed, and the mines' production had been reduced. Inadequate markets for silver and falling prices had "annihilated profits" and produced "inadequate employment for laborers." The paper raised the cry: "Not to England or any other power" need America bow. The plot thickened: England, international Jewish bankers, and eastern big business were all plotting to deprive honest Coloradans of their livelihoods.

After Colorado became the Centennial State and had two senators, Coloradans were not hesitant about writing to Georgetown's former neighbor and now senator, Henry Teller, about their views on the silver question. A sampling of their comments provides the feeling of the day. Fellow lawyer Moses Hallett fumed, "I believe in it [silver]. I am half inclined to think that most of them ['gold chaps'] are first class rascals." "It is the most important question I think of the day, with us in America & especially in Colorado & other silver producing state[s] & territories," maintained miner Richard Irwin, formerly of Georgetown. Not everyone agreed. Joseph Wolff argued eloquently that the government already had enough silver for coinage, so why should it buy more: "Is not our public debt already great enough?"[2]

What Wolff pointed out and the others argued was the fight to regain silver's traditional role in coinage. Demands from the West for "free silver" or "free coinage" attracted support from debtor farmers in the Midwest and, to a lesser degree, in the South. The farmers thought free coinage might result in a "cheaper dollar," thus producing inflation that—they hoped—would generate higher prices for their crops and cattle. That agrarian profit would then expand outward to businesses and railroads, and the entire local community would benefit. The agitators' result was the Bland-Allison Act, passed over the unconvinced President Rutherford Hayes's veto in 1878. The act ordered the treasury to purchase from $2 million to $4 million in silver per month and coin it into silver dollars. Conservative secretaries of the treasury, whether Republican or Democrat, promptly purchased the lower amount.

Furthermore, the government was always ready to redeem silver dollars in gold because silver was "as good as gold." Thus silver was not free, and the silverites' hopes for government support were never really answered. The silver price continued steadily downward throughout the 1880s. That price reflected refined silver from the smelter; mine owners received much less for their raw ore.

The silver issue became more heated and more emotional in the 1880s. A Colorado politician had better stand "four-square" behind silver if he hoped to win an election. Georgetown's and other silver districts' newspapers were filled with articles on the silver question. The state Republican Convention in 1880 addressed the issue, proclaiming that there should be "[e]quality of the gold, silver and paper dollar, each convertible into the other."

Georgetown folks were well aware of the most significant issue they faced. The *Miner* and *Courier* never let the issue rest, as a sampling of quotations attests.

> Every citizen of Colorado ought to feel a lively interest in maintaining this prosperity, and in encouraging the production of silver.

How a citizen of Colorado can entertain other views and still believe himself a friend in the interests of the State, we cannot readily understand.

The enemies of silver in Congress are now making an effort to demonetize that metal.

Demonetization would be a national calamity.

The papers cheered Senator Teller, who had become a great silver spokesman: "The experience of mankind has demonstrated the necessity of the use of both gold and silver as circulating medium; that the destruction of the money faculty of silver is in the interest of a few only and not calculated to benefit the great mass of mankind."[3]

That was where the issue stood as Georgetown entered the 1890s, the third decade of its existence. In Washington that year, eastern senators wanted support for a higher tariff. They were faced by westerners who wanted help for silver and remained lukewarm toward any tariff increase that would benefit the East. The Republicans needed the votes on the tariff of the new western states that had recently joined the Union.

The predictable result came about—political log rolling. Congress passed both the Sherman Silver Purchase Act and the McKinley Tariff. The former required the Treasury to buy 54 million ounces of silver a year, America's estimated annual production. The latter raised duties on manufactured goods to a record high.

Half of the silverites' hopes had been realized: they now had complete government support in purchasing silver. The other half failed to be answered—there was no guaranteed price at 16 to 1. Nor did all the silver need to be coined; the government issued treasury notes redeemable "in gold or silver coin." Would Georgetown benefit from this half a loaf?

The immediate result appeared satisfying. The price of silver jumped 11 cents an ounce, to $1.05. Clear Creek County responded in 1890 by producing the most silver in its history, 1.8 million ounces. Such results did not, however, reflect the true situation of Georgetown mining.

Reports that appeared in the *Engineering and Mining Journal* told a more truthful story. The Pelican-Dives property, "so famous in the early days," showed "more activity" than it had in some time, with lessees working the property. It was hoped this would restore the property "to something like its past greatness." The Silver Age Mine had shut down its concentrating mill. Lessees of the Burleigh Tunnel had done "little with the property this year," but "probably work would be resumed with renewed activity in a short time." The Colorado Central dumps contained eight- to ten-ounce ore, or at least such was rumored, although "[t]ests recently made prove quite conclusively that the popular opinion was wrong." The Pay Rock Consolidated Company stocks' fluctuating share prices "indicate some sort of manipulation," which was "not reassuring."[4]

Little within those comments seemed reassuring. Lessees, old-time producer, restore to past greatness, shut down, little done with property, manipulation, working the dump—those terms did not bode well for Georgetown. They suggested more of an old district trying desperately to stay alive than a new and promising district such as Aspen or Cripple Creek. Optimism or, perhaps more accurately, a long-shot hope had not died, however. Basing its report on news from the county, the *Engineering and Mining Journal* boasted that Clear Creek County would "be the scene of considerable activity in mining this spring [1893]."

The Georgetown mines did not completely fossilize. Some produced quite regularly. Miners there continued to work on the cutting edge of mining technology. The gasoline engine arrived in 1893. "A New Era in Mining Machinery," the *Courier* hailed its introduction at the Troy Mine. As soon as the town gained electrical power, lines were sent to the mines, thus cutting costs and—it was hoped—increasing profits. "It is a step forward in mining, and when thoroughly understood and put into common use will add millions to the output of Colorado mines," predicted the *Georgetown Courier* and other state newspapers.[5]

If the mines showed both their age and a progressive nature, so did Georgetown. Nothing indicated that reality more than the fact that the population had dropped steadily and alarmingly, from 3,294 in 1880 to 2,173 in the 1885 state census, then to 1,927 in 1890. Neighboring Silver Plume, and the county overall, had also declined, whereas Idaho Springs had nearly doubled in population. Unfortunately, Georgetown and its mining represented yesterday, not tomorrow.

Life went on as usual in the community. In some ways Georgetown enjoyed the status of pioneer. Articles about old-timers and pioneers appeared frequently, along with comments about those earlier times. Dick Irwin, for instance, had lived in Georgetown in the 1870s; he had continued prospecting, although now in British Columbia. The *Courier* described him thus: "Irwin is endowed with a prospector's fever that will keep him on the move until time calls a halt."

Electric lights graced the town's streets in June 1891, showing that the community marched not a step behind the modern world, despite being a Colorado pioneer. The *Courier* crowed in October, "[I]t is admitted by experts that Georgetown has the best equipped and most successfully operated electric light plant of any of the smaller cities in the state." In 1893, however, an argument exploded over incandescent lamps versus arc lights, the issue being the high cost of the latter.

Nature showed residents that it, too, could produce electricity. The *Courier* reported: "Electricity darted around last Monday [August 17, 1891] afternoon like bats in the evening, and skipped by many a person all too closely. The electric light wires got well charged and the lamps in the houses kept up a snapping that made the women and children regard them with a certain amount of fear." Additionally, it offered, according to the reporter, "the liveliest display of electricity and the sharpest cracks of thunder that have visited us for many a year."[6]

Problems of age hammered the community in 1889–1890 when its water-works—the pride of 1874—were found to be in disrepair and Georgetown's drinking water was deemed unhealthy. Sadly, several deaths and sicknesses resulted from this situation. The water company revamped and repaired the system, but in 1891 the city acquired the waterworks. The *Courier* supported the purchase, believing profits then would be put back into improving the system. Improvement did come when a city sewer system replaced some of the cesspools scattered about behind homes and buildings. The paper had advocated such a step: "We will never be wholly rid of dangerous diseases in town until we have a good system of sewerage. It is no answer to this to say it is too expensive—there is no expense too great to be undertaken to preserve health."[7] Georgetown was maturing. It was a town whose permanency was assured and whose residents were willing to make improvements to better their quality of life and make their community more attractive.

Georgetown—from its beautiful architecture to its rank-and-file residents—reflected the Victorian Age. Certain attitudes of the era were also clearly reflected in Georgetown. The *Miner* complained in February 1893 that Italian laborers in mines resulted in "an immense amount of money [being taken] out of this state every year." The new immigration of eastern Europeans had started, and many Americans did not like it. Nor did many Coloradans like the federal government's policy of setting aside forest preserves. Georgetown folk signed a petition protesting the establishment of the Clear Creek Forest Reserve: "If carried through [this] would prove a detriment to both domestic and mining purposes."

The newspapers also served as community gadflies. In 1893 the *Miner* wondered why no trees were planted on Arbor Day. The paper suggested that "our city council follow the precedent established by earlier councils to provide trees "to all those who wish them." The editor noted that the schools had held Arbor Day exercises, with songs, readings, essays, and recitations. The editor could also be a curmudgeon, reporting that "Georgetown actually contains several business (?) men that not only refuse to advertise but also refuse to take a home paper."[8]

Georgetown also had problems with some of its youth. When a group fired guns at targets above town and a bullet whizzed through a wheelbarrow on a front lawn, the *Miner* called for city authorities to enforce the ordinance against "discharging guns within the city limits." This "wholesale recklessness must be stopped at once," the editor fumed.

Socially, the town reflected the Victorian Era. Home and church, children and middle-class respectability, and family life centered on Victorian values. A picnic in the mountains for some of Georgetown's young ladies and gentlemen, plus two ministers, caught the attention of a *Courier* reporter in August 1891. When the Methodist church's Epworth League held an installation, it was worthy news. The same was true when the Masonic lodges took "possession of their new quarters in the Fish Block" and the Catholic church redecorated its interior. When some of "our young men" organized a tennis club, it generated newspaper coverage, as did baseball games against heated rivals. When an old "landmark" went down and was replaced by a new, modern structure, that, too, captured *Courier* or *Miner* interest.[9]

What all these items had in common was that they reflected what a settled, respectable community would be doing and achieving in the name of progress and Victorian values. That was the image Georgetown wanted to impart to visitors, investors, and everyone else. Certainly, a red light district existed, with its saloons and other attractions; its existence represented part of male-dominated mining life, besides being good for community business. Very few mining towns had such values, however.

Georgetown and its neighbors had worries, but they had managed to muddle through before and remained optimistic that everything would turn out well. Now, however, they came face-to-face with a disaster that made the past seem tame indeed. In 1893 the hopes and expectations of a generation came crashing down.

In April of that year Colorado collapsed into the worst economic panic and depression in its history. The causes were hauntingly familiar. Agriculture had been depressed for years, inequities within commercial markets aided the East and hurt the rest of the country, and an inflexible monetary system and worldwide economic problems added to the woes. Excessive business and private debt, wild speculation in railroad construction and stock, overexpanison of industry, and an imbalance in foreign trade combined with a growing lack of confidence about the future to spawn a crash.

Before the year was out, all of America had crumpled into a deep depression. Statistics alone do not describe the individual suffering and shock, although they do trace the magnitude. Seventy-four railroads went into bankruptcy. More than 15,000 businesses failed. Hundreds of thousands of people were out of work; some were near starvation. The West and agriculture overall appeared particularly hard-hit, at least in their estimation.

The crash and depression struck Georgetown, like other silver-ribbed communities, especially hard. Not even the local papers could mask the truth of the disaster. On July 1 the *Colorado Miner* acknowledged that "the drop of silver the past week made everything look blue in Georgetown." Comments revealed that "large numbers of people" left this morning "to try and get work elsewhere." They soon learned that none could be found. The Free Coinage Ball at the opera house was only "fairly attended." Coming as it did "in the very midst of our financial disaster," many "were deterred from attending."

The Bank of Clear Creek County, along with scores of others in Colorado, closed its doors. Several months passed before the doors were reopened. Georgetown residents also lost money as banks in Denver and elsewhere, in which they had money deposited, failed. A petition circulated, "signed by all large tax payers," asking the county commissioners to extend the time to pay taxes "until the stringency in the money market is passed." Optimists predicted that it would be sixty to ninety days. That prediction proved dead wrong.

At its July 6 meeting the city council dispensed with the services of the "cart man and the man employed in the park"; reduced the salaries of the marshal, night watchman, and water commissioner; and paid the fire chief only for "actual services rendered." The council also discontinued turning on the streetlights "for the present." The Colorado Central's freight train schedule was reduced, and only one car and a caboose typically appeared. On July 8 the *Miner* fumed about such service, exclaiming "we fear the railroad" has begun "to appreciate our interests too late."

On August 18 the *Miner* reported that "[o]ne of our old time miners," Julian Stanton, had left for Iowa, "where he intends to stay until silver is restored to its rightful place, as one of the money metals of this world." One wonders if he ever returned. Trying to counter the hard times, Georgetown merchants lowered prices. As one explained, "[T]he price of silver and our price of dry goods go down together." The *Miner* had thought, as reported in the July 15, 1893, issue, that mills and smelters should buy silver "at whatever price to allow the mines to reopen." Whether the stores managed to stabilize business remains doubtful; no market for silver existed despite such pleas.[10]

Colorado Bureau of Labor statistics surveyed the "effects of demonetization of silver" in July and August. A grim picture emerged. Of Georgetown it reported 600 men out of work, 200 people having left, forty-five mines closed, and four businesses failed, with the "general feeling, bad." Silver Plume followed suit— 270 were out of employment, three mines had closed, and the general feeling was "serious." Leadville's "gloomy" outlook resulted from ninety mines closing, twenty businesses failing, and 2,500 unemployed—60 percent of whom had left town. On the other hand, neighboring Central City, with its gold, reported no mines closed and a "fair" general feeling. The town claimed about "1,000 miners and prospectors [are] at work in the county. We are not feeling the effects of the slump in silver very much on that account."

Among the comments were one from Clear Creek County: "My reply pertains only to the west half of this county where silver mines are located, which are all closed, and number forty-five large producing properties. It is a very serious question with us now whether they will ever start up again, as the outlook at present is very blue." Georgetown's "bad" feeling compared with other comments from throughout the state, including both mining and farming areas—"depressing, disheartening, deplorable, hopeless, and dead."[11]

The depression that settled in following the crash affected everyone. With no government safety net, people lost their money or received only a percentage back if a bank failed. In Denver alone, twelve banks shut their doors in July. No state or federal relief agencies rode to the rescue of the unemployed, and private and church charities soon ran out of money. People who were out of work flocked to Denver, and Denver put as many as possible on trains and moved them along. Everywhere Georgetown residents looked, there seemed little hope because gold districts like Central City and Cripple Creek could only absorb so many newcomers. All told, in Colorado during the summer of 1893, more than 45,000 people lost their jobs, 435 mines closed, and 377 businesses failed.

Coloradans had never seen a situation this bad, and although the depression hit the entire nation and even the world, it appeared to be worse for them. Not so, said the *Engineering and Mining Journal*: "The Colorado lunatics who have been screaming that Colorado and all other Rocky Mountain mining states will be bankrupted unless the rest of the country will continue to buy their silver at some fancy price have, unfortunately, convinced a great many people that this is true." Other western mining states, it swore, "seem disposed to make the best of the situation." The *Journal*'s analysis did not satisfy many Coloradans. The *Journal* also stated that "Colorado should 'tar and feather' its prophets of evil, and promptly invite them to set out on a quest for a better country, not to return until they had found it."

The *Miner* and *Courier* blasted such eastern attitudes. In this, however, lay the core of the big fight unfolding in Washington. President Grover Cleveland and eastern business and financial establishments chose to blame the crash and the depression on the Sherman Silver Purchase Act.

Blithely, they assumed that the panic had been caused by the fact that in April 1893 the federal gold reserve fell below $100 million for the first time since its creation. Why? Because the Sherman Silver Purchase Act required Uncle Sam to purchase silver and issue dollars and silver certificates redeemable in gold. Although the decline had caused some fears that the government might not be able to maintain the gold standard, that concern completely oversimplified the cause. Cleveland and others supporting economic orthodoxy, in which the gold standard remained the cardinal tenet of modern capitalism, looked for a cause.

They did not have to look far. In their minds the villain—the Sherman Act—needed to be repealed, then all would be well. Cleveland called a special session of Congress that fall, which turned into an epic fight between sound money advocates and silver inflationists. Senator Teller and other western silver men led the emotional battle to save silver. The creditor East stood arrayed against farmer-debtors of the South, Midwest, and trans-Mississippi West. Silver mining interests, fighting for their existence in a changing world, rallied against goldbugs. Nineteenth-century rural America, making a last stand against a future that seemed threatening, struggled against the inevitable.

Georgetown could not sway the outcome, no matter how interested it was in that outcome. Local newspapers kept readers well-informed throughout the year. They did not budge from their silver stand, as these comments reveal.

> Colorado believes in absolute bimetallism because silver is essentially the money of the masses of the people and no universal prosperity can long prevail until it is everywhere remonetized. (January 7, 1893, 1)

> Bankers and gold-bugs are flooding Congress with anti-silver petitions. (February 4, 1893, 1)

> If the great West and South only stand together they can control the finances of this great nation. (May 20, 1893, 2)

> Let the West and South stand solidly for free and unlimited coinage and accept no compromise bill of any kind. (July 15, 1893, 2)

The *Miner* also thought the Sherman Bill "has been a blind from the start. It was simply brought forward to stave off the popular demand for the free coinage of silver."[12]

Georgetown cheered Henry Teller as he fought gamely to save silver. Teller proclaimed before Congress, for example: "If you repeal the purchasing clause of the Sherman act and establish gold monometalism in this country, your legislation will bring distress and disaster and ultimate ruin not only to the people, but the government itself." He knew their fears, he knew their concerns: "Our people are to be left to the tender mercies, to the cruelty and greed of money consolidated in the great commercial centers of this country, reaching out like a great octopus, attacking all the great industries of the country." Would the United States, Teller asked, "surrender the financial independence of this country to England at England's bidding and England's demand?"[13]

Despite local assurance that "the future of silver at the extra session of Congress looks brighter," it was not to be. Congress repealed the Sherman Silver Purchase Act, and the price of silver tumbled—dropping within a year to the range of sixty cents an ounce. By the end of 1893 prospectors had uncovered gold prospects on McClellan Mountain and reported optimistically that a "strong gold belt" ran through that part of the county.[14]

Local gold production went up, and after 1894 silver slipped into a gradual but steady decline. Gold in Clear Creek County would not replace silver, however. Georgetown's mines had been silver mines, and the gold generally lay downstream. The community still held out hopes for free silver at the 16:1 ratio. Their hopes rested on a new political party, the Populists, or People's Party, which had taken up the silver banner. The party, which was extremely popular in Colorado, favored a host of reforms that promised a better tomorrow for Colorado and Americans of all ilk. All eyes looked to 1896, the next presidential election year.

NOTES

1. Henderson, *Mining in Colorado* (Washington, D.C.: Government Printing Office, 1926), 109.

2. *Colorado Miner*, October 7, 1876, 1; Moses Hallett to Teller, February 6, 1878; Richard Irwin to Teller, January 11, 1878; Joseph Wolff to Teller, November 24, 1877, all in the Henry Teller Collection, Denver Public Library.

3. Republican Convention quote from *Rocky Mountain News*, August 27, 1880, 8; first two newspaper quotes from *Colorado Miner*, October 30, 1880, 2; *Georgetown Courier*, March 16, 1882, 3; second two newspaper quotes from April 27, 1882, 1; silver spokesman quote from *Congressional Record*, March 21, 1882, 2100.

4. *Engineering and Mining Journal*, April 5, 1890, 391; August 23, 1890, 225; January 14, 1893, 37.

5. *Engineering and Mining Journal*, April 29, 1893, 429; *Georgetown Courier*, April 14, 1891, 1; July 4, 1891, 2; *Silver Standard* [Silver Plume], January 5, 1895, 1.

6. The previous stories were taken from the *Georgetown Courier*, June 20, 1891, 1; August 22, 1891, 3; October 24, 1891, 3; December 17, 1892, 1, Irwin quote on 3; January 7, 1893, 3; April 29, 1893, 2.

7. *Georgetown Courier*, quote from February 21, 1891, 1; October 24, 1891, 2; Christine Bradley, "Infrastructure Improvements in 19th Century Georgetown," *Journal* 1, 2 (March 1995): 20–24.

8. *Colorado Miner*, quotes from January 28, 1893, 2; February 4, 1893, 2; February 12, 1893, 3; April 22, 1893, 1.

9. *Colorado Miner*, August 22, 1891, 3; quote about Masonic lodges from July 2, 1892, 3; January 14, 1893, 1; quotes about youths from June 17, 1893, 3; *Georgetown Courier*, April 2, 1892, 1; July 2, 1892, 3; July 30, 1892, 3; *Silver Standard* [Silver Plume], June 14, 1890, 1.

10. *Colorado Miner*, July 1, 1893, 3; July 8, 1893, 3; July 15, 1893, 3; August 18, 1893, 3.

11. *Bureau of Labor Statistics Colorado* (Denver: Smith-Brooks, 1893), 1–33.

12. *Engineering and Mining Journal*, August 5, 1893, 131; *Colorado Miner*, January 7, 1893, 1; February 4, 1893, 1; quote about the Sherman Bill from February 11, 1893, 2; May 20, 1893, 2; July 15, 1893, 2.

13. Henry Teller, *Miscellaneous Speeches* (Washington, D.C.: Government Printing Office, 1893), 8, 9, 28, 29, 31.

14. *Colorado Miner*, July 15, 1893, 3; *Engineering and Mining Journal*, November 11, 1893, 503.

Epilogue: 1896

IN THE LIFE OF AN INDIVIDUAL, A COMMUNITY, EVEN A STATE OR NATION, ONLY A FEW times can be called a watershed, or a defining moment. Such was the year 1896. For Georgetown, Colorado, and—to perhaps a lesser degree—the United States, that year marked a turning point beyond which only nostalgia looked back.

No more crucial election faced American voters in the nineteenth century, and only a few since have matched it. At stake were real issues, emotional issues, and issues not totally understood as yet. To Coloradans it appeared simple. A fight to preserve free silver and their way of life hung in the balance, and they became radical in the defense of their conservatism. Tomorrow would be decided by the outcome. Logic went out the window, and emotionalism held sway.

It was East versus West, debtor versus creditor, the common man versus the rich, rural America versus urban America, hardworking ordinary people versus U.S. and foreign bankers, silverites versus goldbugs, "the people versus plutocrats," and old America versus new America in a struggle for the country's heart, soul, and future. In truth, this represented the climax of a struggle that had been going on since the days of the founding fathers, the Jeffersonian versus Hamiltonian concepts of where America should go. The Jeffersonian small farmer made his last stand against the Hamiltonian industrial-urban world. It became, in a sense, the nineteenth versus the twentieth century.

Georgetown was trapped in the midst of this turmoil. Its miners, merchants, and ordinary folk well understood what was at stake for them and their local economy. They stood, as some proclaimed, "at Armageddon" and battled not for the Lord but for their future.

They felt betrayed, ignored, abandoned by the two older political parties. The Republicans had become the party of big business and the gold standard. The minority Democrats, still struggling to gain acceptance after they were "blamed" for helping bring on the Civil War—and the Republicans succeeded in weighing them down with that guilt—did not embrace such radicalism as free silver.

The intriguing Populists had taken up the silver crusade, and Colorado had fallen in love with them. The state gave the party's 1892 presidential candidate, James Weaver, its electoral college vote and elected a Populist governor, Davis Waite. The Populists were intriguing for a variety of reasons. They had unintentionally become a bridge from the old to the new. While upholding the virtues of agrarian, nineteenth-century America, they also looked to the future.

They gathered together farmers, laborers, miners, reformers, women's suffrage advocates, and a host of other groups. They had proclaimed at their first national convention (1891) that "[t]he people are demoralized. The fruits of the toil of millions are boldly stolen to build up colossal fortunes . . . while their possessors dispute the republic and endanger liberty. From the same prolific womb of governmental injustice we breed two great classes—paupers and millionaires." Rural, small-town America contained the virtues that had made America great. Something had changed, however, in this new industrial, urban world—and that was what the Populists set out to fix.

How they planned to change things was fascinating. They championed, among other things, the Australian, or secret, ballot; a graduated income tax; the direct election of U.S. senators; shorter work hours; the initiative and referendum; and, in Colorado at least, women's suffrage. If the idea of "all power to the people" did not seem threatening enough to conservatives, the Populists also advocated that the government should own and operate railroads and telegraph and telephone systems, and they demanded that railroads and corporations return to the federal government lands they had been given and did not actually need.

While championing causes they thought would preserve the old, the Populists laid the basis for the reform movement of the next century—the new. Much of this was off in the future or beyond the immediate needs of Georgetown and its people. For them, it was enough that the Populists backed silver.

The *Georgetown Courier* said as much back in September 1893 when it editorialized, "Very few miners are going to vote to keep political parties in existence that have done and are doing all they can to make paupers of them." The paper had also emphasized that "Gold and Silver are united in the veins in a common destiny which forbids their separation. What gold hath joined together let no man put asunder."[1] Now, in 1896, the newspaper's views were even more outspoken. "FREE AND UNLIMITED COINAGE OF SILVER AND GOLD" shouted the editorial banner.

The Republicans, meeting in St. Louis, nominated Ohioan William McKinley and raised the gold standard high. Amid jeers, western silverites, led by Senator Henry Teller, marched out of the convention. A brief Henry Teller presidential boomlet followed, but in July all eyes turned to Chicago where the Democrats met.

The Democrats were split between the goldbugs and silverites, and the two factions had not come together in a unified position. Then a promising young Nebraska politician, William Jennings Bryan, rose to speak on behalf of the silver plank in the party's platform. With a powerful, cultivated voice that could be heard in every corner of the convention hall, Bryan spoke, "asserting [that] the money question was the paramount issue of the hour." He spoke for the people, he spoke for free silver.

> The man who is employed for wages is as much a business man as his employer; the attorney in a country town is as much a business man as the corporation counsel in a great metropolis . . . the miners who go down a thousand feet into the earth, or climb two thousand feet upon the cliffs, and bring forth from their hiding places the precious metals to be poured into the channels of trade, are as much business men as the few financial magnates who, in a back room, corner the money of the world.
>
> The sympathies of the Democratic party, as shown by the platform, are on the side of the struggling masses who have ever been the foundation of the Democratic party. . . . The Democratic idea, however, has been that if you legislate to make the masses prosperous, their prosperity will find its way up through every class which rests upon them.
>
> You come to us and tell us that the great cities are in favor of the gold standard; we reply that the great cities rest upon our broad and fertile prairies. Burn down your cities and leave our farms and your cities will spring up again as if by magic; but destroy our farms and the grass will grow in the streets of every city in the country.

Moving confidently toward his conclusion, Bryan trumpeted the ideas he had tested and retested in scores of earlier speeches.

> Having behind us the producing masses of this nation and the world, supported by the commercial interests, the laboring interests, and the toilers everywhere, we will answer their demand for a gold standard by saying to them: You shall not press down upon the brow of labor this crown of thorns; you shall not crucify mankind upon a cross of gold.[2]

For a moment "the Boy Orator of the Platte" mesmerized the convention. Its delegates then rose—galvanized by silver—to give their new hero an unmatched, thunderous ovation. The stunned goldbugs could only watch. The convention nominated Bryan for president on the fifth ballot. Coloradans, Georgetownites, and silverites everywhere had found their man.

The Populists, finding that the Democrats had stolen not only the silver issue but most of the rest of their reform program as well, also nominated Bryan. Stubbornly, they selected a different vice presidential candidate. For all practical purposes, however, the Populist Party died that year, although its program lived on.

The *Courier* came out for Bryan and his running mate, Arthur Sewall, in its first issue after their nomination, on July 18. It asserted, "Colorado should give an 'overwhelming majority' to Bryan and Teller, running for reelection. President Bryan will need Senators whom he can rely upon to carry out needed legislation." Optimism, enthusiasm, and expectation crowned the early part of the campaign. The *Courier* observed, "The goldbugs are getting greatly discouraged over the political outlook. It indicates that Bryan will carry every state in the Union."

Picking up on Bryan's speech, the cry, under the nominees' names, was "No Crown of Thorns, No Cross of Gold." By October 31 and the end of the campaign, the press was taking a more realistic position: "We march into the fiercest political fight in 35 years armed with ballots and nothing else."[3] When those ballots were counted in Clear Creek County, Bryan had buried McKinley by a count of 3,342 to 101. The results were similar in the other Colorado mining counties. Alas, the result was not the same across the country.

Bryan lost the election, although Colorado gave him an overwhelming majority of nearly 85 percent of votes cast. Why was he defeated? Silver had not really caught on nationally, other Democratic reform ideas seemed too new and too startling, the Republicans were too well organized and financed, the eastern establishment stayed loyal to the GOP, and in some areas the depression of the 1890s no longer had a grip.

Georgetown residents blamed the defeat on "Wall Street shylocks," money, and big business—clearly showing the divisions within the country at the time. Under the heading "Four More Years of Tory Rule," the *Courier* printed Bryan's letter to the bimetalists in which the hero of the hour—despite his defeat—sounded a cheerful note: "Conscious that millions of loyal hearts are saddened by temporary defeat, I beg to offer a word of hope and encouragement. No cause ever had supporters more brave, earnest and devoted."[4]

Even in defeat, William Jennings Bryan forced Americans to look at themselves and their country. Bryan represented yesterday and tomorrow simultaneously. He championed in 1896 the changes that were coming. He stood for a rural America that was receding slowly into the past, losing both financial and political influence.

Changes were coming for Georgetown as well. Its days as a mining center were disappearing "like a whisper in the wind." In the new century Georgetown would look for new economic pillars, and several generations would pass before those new pillars supported the community. By then, mining would be a relic of a bygone day, romanticized far beyond its reality.

Georgetown's history had followed that of so many of its contemporaries. It had been born, prospered, and declined within the blink of a generation. As short as that might seem, it had lasted longer than most other mining communities that existed while Georgetown was enjoying its mining days. Many had become mere shells of their former glory or stood deserted by everything but memories. Others such as Leadville and Butte had emerged as grim industrial towns. Georgetown, nestled in a beautiful mountain valley, still held promise for the future.

Were the mining days a failure for Georgetown? Emphatically no. The district and the town had helped open, "tame," develop, settle, promote, and mature Colorado and had helped build the foundation that carried the state far into the future. They had been part of the larger settlement of the Rocky Mountains and the West—that epic of westward migration that had started long before in Jamestown, Virginia, and Plymouth, Massachusetts. They had left behind a strong heritage and a legend.

As the years passed, perhaps, the pioneers and settlers who built their homes in the region, mined in the mountains' rocky depths, and lived a full life remembered more than had actually happened. They grew with their legend into something larger then themselves. Future generations would have to separate fact from fiction and gauge what had happened and why. For the moment, it did not matter.

Georgetown residents had brought "civilization" to their remote mountainous valley—from main street stores to church steeples, schoolhouse bells, and proper Victorian homes. They had persevered in overcoming, or at least tempering, obstacles and conditions that would have defeated less determined people. In the end, many of them stayed, their community continued, and together they awaited an unknown future.

Colorado poet Thomas Hornsby Ferril, in his poem "Judging From the Tracks," perfectly captured the spirit of those days of long ago.

> Man and his watchful spirit lately walked
> This misty road . . . at least the man is sure,
> Because he made his tracks so visible,
> As if he must have felt they would endure.[5]

NOTES

1. *Georgetown Courier,* August 5, 1893, 2; September 30, 1893, 2.
2. William J. Bryan, *The Cross of Gold* (Lincoln: University of Nebraska Press, 1996), 17–28.
3. *Georgetown Courier,* July 18, 1896, 1; August 8, 1896, 2; August 29, 1896, 1; October 10, 1896, 1; October 31, 1896, 2.

4. *Georgetown Courier,* November 14, 1896, 1.

5. Robert C. Baron et al., *Thomas Hornsby Ferril and the American West* (Golden: Fulcrum, 1996), 48.

Appendix
Population and Census Tables

1885 Census Breakdown, Georgetown, Clear Creek County

	Birth	%	Ancestry	%
United States	2,026	72.15	1,377	49.04
Ireland	155	5.52	386	13.71
Sweden	136	4.84	197	7.02
Canada	126	4.49	152	5.41
England	109	3.88	207	7.37
Germany	80	2.85	169	6.02
Scotland	41	1.46	85	3.03
Wales	25	0.89	64	2.28
Norway	16	0.57	21	0.75
France	13	0.46	26	0.93
Cornwall	11	0.39	19	0.68
Prussia	11	0.39	16	0.57
China	11	0.39	11	0.39
Italy	10	0.36	14	0.50
Austria	8	0.28	14	0.50
Poland	7	0.26	11	0.39
Switzerland	6	0.21	15	0.53
Russia	5	0.18	17	0.61
Isle of Man	4	0.14	5	0.18
New Zealand	4	0.14	0	0.00
Portugal	1	0.04	2	0.07
Australia	1	0.04	0	0.00
South America	1	0.04	1	0.04
At sea	1	0.04	0	0.00
	2808	100.00	2808	100.00

Note: This table is a compilation of Enumeration District No. 3, 1885 Colorado State Census, as transcribed by the Foothills Genealogical Society in its publication *Colorado Genealogical Chronicles,* Volume XV, Combined 1885 Clear Creek County Colorado State Census. Enumeration District No. 3 includes all of Georgetown as well as some smaller surrounding communities such as Alvarado, Lawson, Downieville, and Dumont.

Population Figures, Colorado and Selected Mineral Counties, 1870–1950

Area	1870	1880	1890	1900	1910	1920	1930	1940	1950
Colorado	39,864	194,327	412,198	541,483	799,044	939,191	1,035,791	1,123,296	1,325,089
Clear Creek	1,596	7,823	7,184	7,082	5,001	2,891	2,155	3,784	3,289
Gilpin	5,490	6,489	5,867	6,690	4,131	1,364	1,212	1,625	850
Lake	522	23,563	14,633	7,016	10,812	11,218	4,899	6,883	6,150
Ouray	—	2,669	6,510	4,731	3,514	2,620	1,784	2,089	2,103
Pitkin	—	—	8,929	7,020	4,566	2,787	1,770	1,836	1,646
Summit	258	5,459	1,906	2,744	2,033	1,724	987	1,754	1,135
Teller	—	—	—	29,002	14,351	6,696	4,141	6,463	2,754

Note: Data from Colorado Department of Local Affairs Website, based on materials from the Demography Section, U.S. Census Bureau, provided by the Colorado State Archives.

Population Figures, Clear Creek County and Municipalities, 1870–1950

Area	Incorp.	1870	1880	1890	1900	1910	1920	1930	1940	1950
County	1861[1]	1,596	7,823	7,184	7,082	5,001	2,891	2,155	3,784	3,289
Empire	1882	150[2]	unknown	134	276	179	105	93	174	228
Georgetown	1868	802	3,294	1,927	1,418	950	703	303	391	329
Idaho Springs	1878	400–500[3]	733	1,388	2,502	2,154	1,192	1,207	2,112	1,769
Silver Plume	1880	"small"[4]	1,200[5]	908	775	460	272	126	139	136

Note: Data from Colorado Department of Local Affairs Website, based on materials from the Demography Section, U.S. Census Bureau, provided by the Colorado State Archives, with additional materials from newspaper sources as noted in later notes.

[1] Clear Creek County was one of Colorado's original counties, created with the establishment of the Territory of Colorado in 1861. The boundaries of the county have never changed.

[2] Figure from *Colorado Miner,* May 19, 1870.

[3] Ibid.

[4] Actually a reference to Brownville, which would have included the area soon to be known as Silver Plume. Probably 50 or less, since towns with populations of 75 or more are given specific numbers. Ibid.

[5] *Colorado Miner,* May 10, 1879. This figure includes the population of Brownville.

Bibliographical Essay

Liston Leyendecker's *The Griffith Family and the Founding of Georgetown* (Boulder: University Press of Colorado, 2001) provides solid insight into the early years of the young mining camp and its founders, as does his earlier publication *Georgetown: Colorado's Silver Queen 1859–1876* (Fort Collins: Centennial, 1977). Several contemporary accounts also render valuable information. Frank Hall's *History of the State of Colorado,* 4 vols. (Chicago: Blakely, 1889–1895) presents the state's history through his recollections as one who lived in Central City through the boom years and then moved into state government, where he dealt with most Georgetown notables throughout the time frame of his publication. The Clear Creek County section of O. L. Baskin's *History of Clear Creek and Boulder Valleys* (Chicago: O. L. Baskin, 1880), written by local newspaperman Aaron Frost, includes a general history of the towns and mining industry, as well as biographical information from those interested in submitting material for publication.

Three local newspapers (in order of their appearance on the scene)—the *Colorado Miner* [Georgetown], the *Georgetown Courier,* and the *Silver Standard* [Silver Plume]—recount the events of the day with different perspectives and political leanings. Those wishing to focus on primary research will find the manuscript collections at the Archives, Norlin Library, University of Colorado in Boulder, especially helpful (particularly the Jerome B. Chaffee, Harper M. Orahood, and Henry Teller collections), as well as the State Historical Society Library in Denver (Edward O. Wolcott, Nathaniel P. Hill, Frederick W. Pitkin, and Jesse Randall collections, among others).

John Horner's *Silver Town* (Caldwell, Idaho: Caxton, 1950) used extensive newspaper references along with insights gathered from many of Georgetown's older residents, whose firsthand accounts gave the author the feel of days gone by.

Several statewide religious studies document the early years of the denominations in Georgetown: Thomas J. Noel, *Colorado Catholicism and the Archdiocese of Denver, 1857–1989* (Boulder: University Press of Colorado, 1989); Allen D. Breck, *The Episcopal Church in Colorado, 1860–1963* (Denver: Big Mountain, 1963); Alton Templin, Allen D. Breck, and Martin Rist, eds., *The Methodist, Evangelical, and United Bretheren Churches in the Rockies, 1850–1976* (Denver: Rocky Mountain Conference of the United Methodist Church, 1977); Walter S. Hopkins and Virginia Greene Millikin, *The Bible and the Gold Rush: A Century of Congregationalism in Colorado* (Denver: Big Mountain, 1962); Ida Libert Uchill, *Pioneers, Peddlers, and Tsadikim* (Boulder: Quality Line, 1957); Allen D. Breck, *A Centennial History of the Jews of Colorado, 1859–1959* (Denver: Hirschfeld, 1960). An excellent local study was done by Cynthia C. Wadsworth (now Cynthia C. Neely), "Steadfast in Faith: A Book of Memories, 1874–1974" (Georgetown, Colo.: First United Presbyterian Church, 1974). Rev. Isaac H. Beardsley's memoirs, *Echoes From Peak and Plain: Or, Tales of Life, War, Travel and Colorado Methodism* (Cincinnati: Curtis and Jennings, 1898), provide an excellent firsthand account of the early days of the Methodist church in Colorado.

James Wright's *Politics of Populism: Dissent in Colorado* (New Haven: Yale University Press, 1974) covers the economic and social ends of the Populist movement in the last half of the nineteenth century, including several references to Georgetown. *The Making of Western Labor Radicalism: Denver's Organized Workers, 1878–1905* by David Brundage (Urbana: University of Illinois Press, 1994) sets the stage for the issues and emotions in Georgetown in the late nineteenth and early twentieth centuries.

Duane Smith's *Rocky Mountain Mining Camps: The Urban Frontier* (Bloomington: Indiana University Press, 1967) and *The Birth of Colorado: A Civil War Perspective* (Norman: University of Oklahoma Press, 1989) give the reader an understanding of the life and times of nineteenth-century Colorado, as do his biographies *Horace Tabor, His Life and the Legend* (Boulder: Colorado Associated University Press, 1973) and *Henry M. Teller, Colorado's Grand Old Man* (Boulder: University Press of Colorado, 2002). Leanne Sander's Ph.D. dissertation, " 'The Men All Died of Mining Disease': Women and Families in the Industrial Mining Environment of Upper Clear Creek, Colorado, 1870–

1900" (University of Colorado at Boulder, 1990), examines local mining communities through the lives and experiences of local Irish families.

Two of the best railroad histories include Cornelius Hauck's *Narrow Gauge to Central and Silver Plume: Colorado Rail Annual Number 10* (Golden: Colorado Railroad Museum, 1972) and *Georgetown and the Loop* (Denver: Rocky Mountain Railroad Club, 1988) by P. R. "Bob" Griswold, Richard H. Kindig, and Cynthia Trombly. Dan Abbot's *Stairway to the Stars, Colorado's Argentine Central Railway* (Fort Collins: Centennial, 1977) covers a later time period but is thoroughly researched and well illustrated.

Erl and Carrie Scott Ellis document the history of the people and lands between Silver Plume and the Continental Divide in *The Saga of Upper Clear Creek* (Frederick, Colo.: Jende-Hagan, 1983), an invaluable reference guide. Louise Harrison's *Empire and Berthoud Pass* (Denver: Big Mountain, 1964) provides extensive materials related to both Georgetown and Empire as well as the big road to the west, an integral part of Georgetown's early history. David S. Digerness presents the history of Georgetown, Empire, and Idaho Springs through the memoirs and photos of his grandfather, local editor Jesse Randall, in *The Mineral Belt, Volume III—Georgetown-Mining–Colorado Central Railroad* (Silverton, Colo.: Sundance, 1982).

Several booklets further illustrate the history of the town and memories of longtime residents, including Polly Chandler's booklets *This Is Georgetown* (Georgetown, Colo.: self-published, 1972) and *The Men Who Lived on Hope* (reprinted from *Empire Magazine* of the *Denver Post*, 1979). Also W. D. Copeland, *One Man's Georgetown* (Georgetown, Colo.: self-published, 1973). Ellen Riddle's biography of Louis Dupuy, *Louis Dupuy and His Souvenir of France* (Mission Viejo, Calif.: John Ray Riddle, 1985), and Christine Bradley's *William A. Hamill: The Gentleman From Clear Creek* (Fort Collins: Colorado State University Cooperative Extension Service, Historical Bulletin 2, 1977) focus on key individuals within the town's past.

Over the years, Georgetown's mining has not received the scholarly attention it deserves. The nineteenth century, however, does provide some fascinating glimpses of this era. Ovando Hollister wrote two books that touch on Georgetown: *The Silver Mines of Colorado* (Central City, Colo.: Collier and Hall, 1867) and *The Mines of Colorado* (Springfield, Mass.: Samuel Bowles, 1867). *The Mines of Clear Creek County, Colorado* (Denver: Times-Steam, 1876) by Samuel Cushman focuses completely on the district. U.S. Commissioner of Mining Statistics Rossiter Raymond published eight volumes (1869–1877) that have chapters on Colorado and sections on Georgetown. James W. Taylor, *Report on the Mineral Resources East of the Rocky Mountains* (Washington, D.C.: Government Printing Office, 1868) preceded Raymond. Frank Fossett's various editions of *Colorado* (1870–1880) contain a wide variety of information: see *Colorado: A Historical Descriptive and Statistical Work on the Rocky Mountain Gold and Silver Mining Region* (Denver: Daily Tribune

Steam Printing House, 1876) and *Colorado; Its Gold and Silver Mines, Farms and Stock Ranges and Health and Pleasure Resorts* (New York: C. G. Crawford, 1879–1880).

More recent publications that touch on aspects of Georgetown mining include Jay Fell, *Ores to Metals* (Lincoln: University of Nebraska Press, 1979); Charles Henderson, *Mining in Colorado* (Washington, D.C.: Government Printing Office, 1926); T. A. Rickard, *Interviews With Mining Engineers* (San Francisco: Mining and Scientific Press, 1922); Ronald Brown, *Hard-Rock Miners* (College Station: Texas A&M Press, 1977); and Clark C. Spence, *British Investments and the American Mining Frontier* (Ithaca: Cornell University Press, 1958) and *Mining Engineers and the American West* (New Haven: Yale University Press, 1970).

Liston Leyendecker's study of the Dives-Pelican feud above Silver Plume is an excellent review of mining law and the problems that came with success: "The Pelican-Dives Feud," *Essays and Monographs in Colorado History* (Denver: Colorado Historical Society, 1985).

Two of the most indispensable sources of Georgetown mining history are newspapers and mining journals. The *Engineering and Mining Journal* (originally *American Journal of Mining*), 1859–1896, contains a multitude of items on Georgetown and Colorado mining. It also places the Georgetown District within a national scope. The Georgetown newspapers contain even more information, particularly in their year-end mining summaries and mining columns, but they must be weighed carefully against local promotion and pride. Denver's *Rocky Mountain News* is a valuable source of information on a variety of Georgetown topics. Central City newspapers carry Georgetown items as well, as do, to a lesser degree, some of the other nearby local papers.

Individual collections in various archives and historical societies in Colorado and elsewhere contain information on Georgetown individuals and mining. The reader is encouraged to examine the footnotes for these sources.

Index

working, 152, 153, 222, 225, 226, 228, 275, 276; management of by Coloradans, 225; naming of, 63; owners of, 132; problems with, 123; promotion of, 67–8, 125; prospecting and development of diminished, 222; prospects of, 126; rail tramways and, 228; "tunnel craze" in, 68, 128–9; tunnels, 222, 226; unprofitability of Georgetown's gold, 14; unsystematic development of, 133; value of production of, 151–3, 226, 272

mining, 97, 165, *166–80,* 270; in Clear Creek County, 64; Colorado's transition to lode, 22; costs of, 272; dangers of, 77, 228; decline of in Georgetown, 286–7; early lode mining in Georgetown, 10–11; effects of Depression of 1893 upon, 279; end of Georgetown's gold era, 14; environmental consequences of, 132–3, 251; nature of for silver, 62; number employed in, 140; photographs of underground, *178;* problems of, 65; processes of, 139; reduced transience in, 146; revival of silver in 1870s, 127; speculators and, 223; technological revolution in, 127; technologies of, 126, 276; tunnel driving, 67–8, 127; tunnels, advantages of, 128–9; tunnels, inadequacies of, 128–9; underground, 139; and use of public lands, 251. *See also* gold; silver; silver ores

Mining and Business Men's Association, 228
mining claims, location of, 16n34
mining companies, failures of, 65
mining disputes, violence in, 135–7
mining districts: development of laws regulating, 9; government of, 9; map of those surrounding Griffith Mining District, 27; organization of, 17
mining engineers, 139
mining investment. *See* investment
mining law: models of for Colorado's districts, 17; Spanish, 17; U.S. of 1866, 18; U.S. of 1872, 133
mining litigation, 130; and Georgetown's reputation, 134; Pelican and Dives conflict, 133–7; over mines and property in Georgetown, 32
Mines of Colorado (Hollister), 40
Mint (saloon), 156
Missouri, 140
Mitchell, Rev. D. H., 85
Moffat, David H., 102
Montana, 123–4
Montana Mining District, 9
Monte Bros. (store), 33
Monti, Joshua, 86, 94
Monti's Hall, 84

Morning Star (Leadville mine), 220
Morrell, M. T., 250
Morris, James R., 158
Morrison, Robert S., 75
Mount Alpine Gold Company, 13, 14, 28, 29, 32, 33, 65
Mount McClellan. *See* McClellan Mountain
municipal government. *See* government
Murphy Hall, 156
Muscovite (mine), 63

Nan Smith Trail, 24
Napheys, B. F., 147
Nash, Harry H., 93, *121*
National Mining and Industrial Exposition (Denver), 226
National Mining Law of 1872. *See* mining law
Native Americans. *See* Indians
Nelander, Edward, 190
Nelson, Albert, residence of, *202*
Nelson, Ed, *254*
Nevada, 61, 63, 123, 127, 133
New Boston (mine), 63
New England: cultural replication of in Georgetown, 33, 43; former residents of in Georgetown, 34; Georgetown's government modeled on local government in, 34–5; New England, influences of upon Georgetown, 85, 88; origins and characteristics of Georgetown's developers, 29, 31
New Hampshire (mine), 63
New Mexico, 5
New Year's Day, 158
New York Times, 127
newspaper, 83, 95, 163–4, 200–201; belittling of new districts, 224; boosterism of, 195, 255; coverage of Georgetown, 222; mine promotion by, 67; town promotion by, 67. *See also Colorado Miner; Georgetown Courier*
Nichols, George L., 10, 12
Nichols, W. H. J., 59
Ninth Street, *108,* 184, 206, 243, 246
No Name (mine), 63
Nominee (lode), 24
North Carolina, gold rush in, xiii
North Clear Creek. *See* Clear Creek, North Fork
North Park, Colorado, 4
Nott, Silas, 94; residence of, 248
Noxon, A. M., 26
Nuckolls, Stephen F., 13, 14, 28
Number 5 Lode, 78
Nyanza (mine), 63